THE MATHEMATICS OF OZ

Mental Gymnastics from Beyond the Edge

Grab a pencil. Relax. Then take off on a mind-boggling journey to the ultimate frontier of math, mind, and meaning as acclaimed author Dr. Clifford Pickover, Dorothy, and Dr. Oz explore some of the oddest and quirkiest highways and byways of the numerically obsessed. Prepare yourself for a shattering odyssey as *The Mathematics of Oz* unlocks the doors of your imagination. The thought-provoking mysteries, puzzles, and problems range from zebra numbers and circular primes to Legion's number – a number so big that it makes a trillion pale in comparison. The strange mazes, bizarre consequences, and dizzying arrays of logic problems will entertain people at all levels of mathematical sophistication.

The tests devised by enigmatic Dr. Oz to assess human intelligence will tease the brain of even the most avid puzzle fan. Test your wits on a host of mathematical topics: geometry and mazes, sequences, series, sets, arrangements, probability and misdirection, number theory, arithmetic, and even several problems dealing with the physical world. With numerous illustrations, this is an original, fun-filled, and thoroughly unique introduction to numbers and their role in creativity, computers, games, practical research, and absurd adventures that teeter on the edge of logic and insanity. *The Mathematics of Oz* will have you squirming in frustration and begging for more.

Clifford A. Pickover received his Ph.D. from Yale University and is the author of over twenty highly acclaimed books on topics such as computers and creativity, art, mathematics, black holes, human behavior and intelligence, time travel, alien life, and science fiction. His Web site, www.pickover.com, has received over a half-million visits.

Works by Clifford A. Pickover

The Alien IQ Test
Black Holes: A Traveler's Guide
Chaos and Fractals
Chaos in Wonderland
Computers, Pattern, Chaos, and Beauty
Cryptorunes: Codes and Secret Writing
Dreaming the Future
Future Health: Computers and Medicine in the 21st Century
Fractal Horizons: The Future Use of Fractals
Frontiers of Scientific Visualization (with Stu Tewksbury)
The Girl Who Gave Birth to Rabbits
Keys to Infinity
The Loom of God
Mazes for the Mind: Computers and the Unexpected
Mind-Bending Visual Puzzles (calendars and card sets)
The Paradox of God and the Science of Omniscience
The Pattern Book: Fractals, Art, and Nature
The Science of Aliens
Spider Legs (with Piers Anthony)
Spiral Symmetry (with Istvan Hargittai)
Strange Brains and Genius
The Stars of Heaven
Surfing through Hyperspace
Time: A Traveler's Guide
Visions of the Future
Visualizing Biological Information
Wonders of Numbers
The Zen of Magic Squares, Circles, and Stars

THE
MATHEMATICS
OF *OZ*

*Mental Gymnastics from
Beyond the Edge*

CLIFFORD A. PICKOVER

CAMBRIDGE
UNIVERSITY PRESS

PUBLISHED BY THE PRESS SYNDICATE OF THE UNIVERSITY OF CAMBRIDGE
The Pitt Building, Trumpington Street, Cambridge, United Kingdom

CAMBRIDGE UNIVERSITY PRESS
The Edinburgh Building, Cambridge CB2 2RU, UK
40 West 20th Street, New York, NY 10011–4211, USA
477 Williamstown Road, Port Melbourne, VIC 3207, Australia
Ruiz de Alarcón 13, 28014 Madrid, Spain
Dock House, The Waterfront, Cape Town 8001, South Africa

http://www.cambridge.org

Typefaces Poppl-Pontifex 9.5/14 pt. with Poppl-Laudatio and Serifa
System QuarkXPress™ [MG]

A catalog record for this book is available from the British Library.

Library of Congress Cataloging in Publication Data
Pickover, Clifford A.
The mathematics of Oz: mental gymnastics from beyond the edge / Clifford A.
Pickover.
p. cm.
Includes bibliographical references and index.
ISBN 0-521-01678-9
1. Mathematical recreations.   I. Title.
QA95.P528 2002
793.7′4–dc21

2002023346

ISBN 0 521 01678 9 hardback

This book is dedicated to my uncle, Dr. Bruce Pickover,
who stimulated my early interest in mathematical and other puzzles.

"Can't you give me brains?" asked the Scarecrow.

"You don't need them. You are learning something every day. A baby has brains, but it doesn't know much. Experience is the only thing that brings knowledge, and the longer you are on earth the more experience you are sure to get."

"That may all be true," said the Scarecrow, "but I shall be very unhappy unless you give me brains."

> — The Scarecrow conversing with Oz, in L. Frank Baum's *The Wonderful Wizard of Oz*

Contents

Travel Guide

He calmly rode on, leaving it to his horse's discretion to go which way it pleased, firmly believing that in this consisted the very essence of adventures.
— Miguel de Cervantes, *Don Quixote*

The road through this book is chaotic and takes many turns in order to surprise and delight you. However, if you wish to take your hovercraft and jump between mathematical puzzles of similar kinds, the following guide should help.

🏠† ✍ **You are here**
- **Geometry** (Chapters 1, 3, 8, 14, 16, 17, 18, 23, 47, 50, 54, 55, 58, 61, 65, 84, 88, 96, 103, 104, 106)
- **Mazelike Problems** (Chapters 6, 12, 13, 15, 22, 24, 36, 46, 49, 52, 60, 83, 87, 97, 98, 101, 102)
- **Sequences, Series, Sets, and Arrangements** (Chapters 2, 4, 5, 9, 11, 25, 26, 30, 34, 37, 38, 41, 48, 50, 56, 59, 63, 64, 66, 69, 71, 72, 73, 74, 79, 80, 82, 85, 86, 89, 92, 93, 107)
- **Physical World** (Chapters 1, 3, 7, 40, 44, 45, 78, 102)
- **Probability and Misdirection** (Chapters 8, 14, 17, 19, 27, 31, 32, 51, 70, 81, 90)
- **Number Theory and Arithmetic** (Chapters 9, 10, 20, 21, 27, 33, 35, 37, 39, 42, 43, 45, 47, 53, 57, 62, 67, 68, 76, 77, 82, 91, 99, 100, 105, 108)
End here ☞ †🏠
(Freedom for Dorothy and Ultimate Reader Enlightenment)

B.C. MANSFIELD

Preface

"If you only had brains in your head you would be as good a man as any of them, and a better man than some of them. Brains are the only things worth having in this world, no matter whether one is a crow or a man."
  — Crows talking to the Scarecrow in *The Wonderful Wizard of Oz*

Oz is a metaphor for mystery. Oz is a state of mind. Oz is a parallel universe that may exist side by side with our own in some ghostly fashion.

Published in 1900 by L. Frank Baum, *The Wonderful Wizard of Oz* starred young Dorothy of Kansas along with a magical array of characters ranging from a Scarecrow and a Tin Woodman to a phony Wizard who used magic to help Dorothy come to realize that there is no place like home.

In *The Mathematics of Oz* Dorothy is certainly far from home. Abducted by mathematically obsessed aliens, Dorothy tries to solve a baffling array of brainteasers that often center around numbers and mathematics. The aliens' obsession with mathematics probably sounds silly to many of you, but numerical challenges are a great way to transcend space and time. Mathematics is a universal language, and numbers might be our first means of communication with intelligent alien races.

Dorothy, Dr. Oz (her abductor), and Mr. Plex (Oz's assistant) have a limited attention span, and they don't want you readers to wade through pages of background before getting to the essential ingredients. To avoid this problem, each chapter in this book is just a few pages in length. One advantage of this format is that you can jump right in to experiment and have fun. The book is not intended for mathematicians looking for formal mathematical explanations; however, additional material can be found in the "Further Exploring" and "Further Reading" sections.

Prepare yourself for a strange journey as *The Mathematics of Oz* unlocks the doors of your imagination. The mysteries, puzzles, and problems range from building a yellow-brick road that crosses America, to zebra numbers and circular primes, to Legion's number – a number so big that it makes a trillion pale in comparison – to "The Problem of the Bones," a fiendishly difficult mathematical problem involving probability and shattering of leg bones.

Grab a pencil. Do not fear. Some of the topics in the book may appear to be curiosities, with little practical application or purpose. However, I have found these experiments to be useful and educational – as have the many students, educators, and scientists who have written to me. Throughout history, experiments, ideas and conclusions originating in the play of the mind have found striking and unexpected practical applications. Or as mathematician Gottfried Wilhelm Leibniz once said, "Les hommes ne sont jamais plus ingénieux que dans l'invention des jeux." ("Men are never more ingenious than when they are inventing games.")

✳   ✳   ✳

This book is for anyone who wants to enter new mental worlds. If you are a teacher, you may want to use the mathematical brain teasers to stimulate students. Have them design their own puzzles similar to the ones in this book. Computer programmers may want to create or solve similar puzzles using a computer, although a computer is definitely not necessary to attack and solve the problems in this book.

To help you assess your level of performance during your journey through this book, I have assigned difficulty ratings to the various puzzles:

✳   Challenging
✳✳   Very challenging
✳✳✳   Extremely difficult
✳✳✳✳ Outrageously difficult: probably impossible for Dorothy and other *Homo sapiens* to solve.

To retain the playful spirit of the book, and its sense of crazy adventure, puzzles with different difficulty levels are scattered randomly through the book – as if the puzzles had been tossed about by a tornado. Browse from the mathematical smorgasbord and feed your mind.

Acknowledgments

Every now and again one comes across an astounding result that closely relates two foreign objects which seem to have nothing in common. Who would suspect, for example, that on the average, the number of ways of expressing a positive integer n as a sum of two integral squares, $x^2 + y^2 = n$, is π.

— Ross Honsberger, *Mathematical Gems III*

I thank Brian C. Mansfield for his wonderful cartoon diagrams, used throughout the book. Over the years, Brian has been helpful beyond compare.

Numerous people have provided very useful feedback and information for the solutions to my puzzles: Dennis Gordon, Robert Stong, David T. Blackston, Dennis Yelle, Balakumar Jothimohan Balasubramaniam, Ilan Mayer, Ed Murphy, Jim Gillogly, Dan Tilque, Bill Ryan, James Van Buskirk, "R.E.S.," Dennis Gordon, Dharmashankar Subramanian, Richard Heathfield, Al Zimmerman, Risto Lankinen, Seth Breidbart, Darrell Plank, David A. Karr, Jason Earls, Ken Inoue, and others.

I thank Samuel Marcius for the symbol 💀, which represents Mr. Plex, and for other alienlike symbols used in this book. The animal font is a freeware font by Alan Carr. Ann Stretton designed the font that contains symbols such as ✳. Symbols like ♪ are part of freeware from Omega Font Labs. Michael Lee and Josh Dixon designed the font that looks like ⑾ⅉ:ʔ‼.

All Oz quotations come from L. Frank Baum's classic novels, *The Wizard of Oz* (originally published in 1900 as *The Wonderful Wizard of Oz*), *The Land of Oz* (originally published in 1904 as *The Marvelous Land of Oz*), *Ozma of Oz* (1907), *Dorothy and the Wizard in Oz* (1908), *The Road to Oz* (1909), *The Emerald City of Oz* (1910), *The Patchwork Girl of Oz* (1913), *Tik-Tok of Oz* (1914), *The Scarecrow of Oz* (1915), *Rinkitink in Oz* (1916), *The Lost Princess of Oz* (1917), *The Tin Woodman of Oz* (1918), *The*

Magic of Oz (1919), and *Glinda of Oz* (1920). For more information on Oz, see Eric P. Gjovaag's "Oz" Web site: http://www.eskimo.com/~tiktok/.

Note: As many readers are aware, Internet Web sites come and go. Sometimes they change addresses or completely disappear. The Web site addresses (URLs) listed in this book provided valuable background information when this book was written. You can, of course, find numerous other sites relating to many of the mathematical puzzles and topics in this book by using search engines such as www.google.com.

Introduction

B.C. MANSFIELD

"Of course I cannot understand it," he said. "If your heads were stuffed with straw, like mine, you would probably all live in the beautiful places, and then Kansas would have no people at all. It is fortunate for Kansas that you have brains."

— The Scarecrow to Dorothy and friends, *The Wonderful Wizard of Oz*

Dorothy lives in the midst of the great Kansas prairies, in the early twenty-first century, with Uncle Henry, a farmer, and Aunt Em, the farmer's wife. One day, while taking a brisk stroll through the prairie grass, Dorothy comes upon a dark monolith protruding from the ground.

Dorothy looks back at her little dog, Toto. "What on Earth could that be, Toto?"

Toto barks and sniffs at the strange object that sits in a capsule of translucent white light as bright as fog on an autumn morning. The light transforms the monolith into an object of great dignity and unbearable beauty.

"Toto, stay back!" Dorothy yells.

Toto approaches the monolith

But it is too late. The moment Toto touches the monolith with his front left paw, the monolith begins to tremble. There is a low scratching sound, like a broom on wet cement.

A large squidlike creature oozes from one of the crevices in the monolith. His tentacles are covered with bubbly skin that occasionally pulsates, its little balloons inflating and deflating. His eyes are iridescent balls, the size of avocados, with hard, crystalline corneas.

The alien looks at Dorothy. "My name is Dr. Oz. Do not fear me."

Dorothy clenches her fists and steps back.

"Dorothy, come with me. I have a test for you. If you pass the test, you may return to your beloved Aunt Em and Uncle Henry."

"Toto," Dorothy whispers, "this must be a dream." Her heart thumps like an anxious conga-drum player. She starts to run.

"Wait," Dr. Oz yells. He aims a wet rectangular device at her left thigh, and within seconds they are both transported to a parallel universe.

Dr. Oz motions for Dorothy to follow him. "Dorothy, ahead of us is a secret testing facility located in a pasture near Lebanon, Kansas, close to the Nebraska border."

"Get away from me, you fiend!"

Dr. Oz grins as he watches Dorothy peer at the octagonal building's high walls the color of emeralds. "We call this place Oz."

Toto remains motionless, as if he is frozen solid. The hairs on his body stand at right angles with respect to his cold flesh.

"My dog – what have you done to him?"

With a wave of Dr. Oz's tentacles, Toto reanimates and jumps into Dorothy's outstretched arms.

In a moment, Dorothy stands on the threshold of a large ultra-modern building and looks back at the countryside. From the far north she hears a low wail of the wind, and she can see the long grass bowed in waves before the coming storm. A sharp whistling fills the air. Is it really the wind, or is the motion caused by some hidden creature?

Dr. Oz extends one of his massive tentacles, touches Dorothy's delicate shoulder, and then gestures for Dorothy to come closer. The tests begin.

B.C. MANSFIELD

1 The Yellow-Brick Road

"The road to the City of Emeralds is paved with yellow brick," said the Witch, "so you cannot miss it. When you get to Oz do not be afraid of him, but tell your story and ask him to help you."
— The Witch of the North, *The Wonderful Wizard of Oz*

Dorothy is in the bowels of the Oz testing facility. Luckily, Toto is still in his normal animated state as he sniffs at the writhing tentacles of Dr. Oz. Oz's appendages remind Dorothy of a bed of twisting snakes.

"Dorothy, do not worry. We will not harm you. We just wish to test the reasoning powers of your species, *Homo sapiens.*"

"Homo, what?"

Dr. Oz waves one of his tentacles. "Never mind that. If you can solve all my puzzles, I will return you to your Kansas farm. If you like, you can occasionally solicit help from your friends using the microphone and transmitter I have implanted in the palm of your left hand."

"You sinister monster. How can you have done this to me?"

"Do not worry. I can easily remove the device. Now listen carefully. Here is your first question. Imagine a yellow-brick road stretching from the East Coast to the West Coast of America."

On a viewscreen there appears a picture of a road stretching across America:

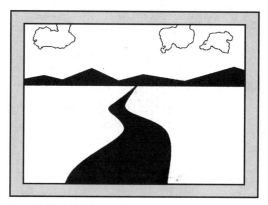

Long yellow-brick road

"Dorothy, I want you to estimate the number of bricks in such a road and tell me how you arrived at the estimate."

Dr. Oz hands Dorothy a pencil and paper and map of the world:

Map of the world

Dorothy focuses on America and sketches a long, straight road across the United States of America while estimating the road's length and width.

Dr. Oz nods. "Dorothy, I'm glad you have calmed down. Now, I also want you to think about what impressive structures you might build with the number of bricks needed to form the road."

"Like what?" Dorothy says as Toto lifts his leg and relieves himself on Dr. Oz.

Dr. Oz pauses. "Do you think your transcontinental road requires millions of bricks, billions of bricks, trillions of bricks, or even more bricks? *Do you think that a thousand Great Pyramids of Egypt would be sufficient to hold the bricks?*"

"Dr. Oz, stop already with these questions! Let me think!"

Dorothy begins to consider the difficult problem as she runs her fingers lingeringly through Toto's hair. A fishy odor begins to fill Dr. Oz's laboratory, quickly followed by the smell of absinthe. On the ceiling, Dorothy notices an octopuslike robot dangling several jointed arms into aquaria teeming with tiny squids. Perhaps the robot assists with feeding the aquarium inhabitants, she thought.

Dorothy looks at her dog. "Toto, I have a feeling we're not in Kansas anymore."

Difficulty Level: ✺

2 Animal Array

"My life has been so short that I really know nothing whatever. I was only made day before yesterday. What happened in the world before that time is all unknown to me."

— Scarecrow, *The Wonderful Wizard of Oz*

Dr. Oz and Dorothy have left the Oz testing facility and are strolling through a small Kansas farming village where Mennonites and other cultures coexist. Dorothy sits down on a bench in front of the Sunflower Buggy Shop.

Dr. Oz turns to Dorothy. "I've placed a cloaking field around my body so that others will see me as human."

"When will you let me go back to Aunt Em and Uncle Henry?" She is upset at herself for not having immediately contacted them using her implanted communicator – which has now mysteriously vanished, leaving no trace of a scar.

"How often do we need to discuss this? First I must test you. And don't try to escape, or else I will keep Toto as a pet for myself."

The owner of the shop, a tall bearded man, sits next to Dorothy and Dr. Oz and begins to reminisce about the Amish life-style and the need for buggies that he constructs and repairs. Dorothy is unsure of the century. Evidently Dr. Oz can transport her through time, space, and alternative realities.

Dorothy listens as Dr. Oz has an interesting conversation with the bearded man on the philosophy of life. The owner practices love, peace, and the pursuit of wisdom by self-discipline. Their discussion at an end, Dr. Oz promptly devours him.

"Gross!" shouts Dorothy. "You're evil!"

"Don't worry. He was actually another alien trochophore, like myself, masquerading as a human. Now, pay attention. Here is your next puzzle."

Dr. Oz begins to sketch on the bench. "Fill in the empty tile with the correct missing two animal symbols."

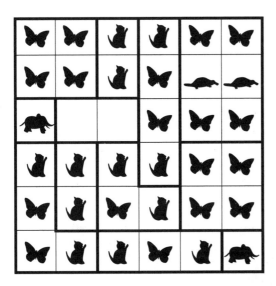

The brilliant Dorothy promptly gives Dr. Oz the answer. Can you?

3 An Experiment with Kansas

B.C. MANSFIELD

"I do not know where Kansas is, for I have never heard that country mentioned before. But tell me, is it a civilized country?"
— The Witch of the North, *The Wonderful Wizard of Oz*

"Follow me," Dr. Oz says as he leads Dorothy and Toto deeper into the bowels of the Oz testing facility. In several minutes they approach a great green door, on which is inscribed the cryptic symbols:

```
ᛈᛁᚱᚾᛁᛁ ᛏᛁ ᚻᛏ. ᛏ ᛪᛁᚲᛏ ᛤᛁᚾ ᚻᛁᛤᚫ ᛤᛁᚾᚱ
'ᛏᚫᛤ. ᛤᛁᚾᚱ ᚻᚠ ᛏᛁᛏᛁ ᚦ ᚻᛏᚾᚻᚠᚫ ᚠ
ᚱᛤᛒᛁᛏ⁛ ᛪᛁᛈᚠᚠᚱ⁛ ᛤᛁᚾ ᛈᛏᚠ ᛏᚻᚠᛁᚱ ᛒᚠ ᚠᛒᚠᛏ
ᛏᛁ ᛏᚻᚠ ᛏᛪᚠᛁ ᛒᚻᛁᚾᛁᛁ ᛪᛁ ᛒᛁᛪᚠᛈᛁ ᚻᛪᚾᛏᚠᛤ
ᚻᛁᚠ ᚠ ᚱᚻᚠ ᚻᚠ. ᛈᚠ ᛪᚠᚠᛏ ᛏᛁᚲᚠᛁᛏᛁᛏ ᛈᛏᚻᛁ
ᚾᚠᛁᚱᚻᛁ ᚾ ᛏᛁᛏᛁ ᛏᛪᚠᛏ ᛈᚠ ᚾᚠᛏ ᛁᛁᚻᛏᛏᛁᚱ
ᛤᛁᚾᚱ ᚻᚠᛁᚱᛤ ᛁᛁᛈᛏ.
```

Dorothy puts her hands on her hips. "What's that supposed to mean?"

"Don't worry about it. It's a secret."

"Why are you really testing me?"

"We intend to colonize Earth if your intelligence is below an acceptable level."

"That sounds cruel," Dorothy says as she cradles little Toto in her arms.

"Don't you want to know more about us?"

"Yes, what exactly is a troposphere? Where do you come from?"

Dr. Oz pauses and points a tentacle at Dorothy. "We are called *trochophores,* and we come from the star Betelgeuse in the constellation Orion. Betelgeuse is a red-giant superstar that could fit a million of your Suns inside of it. But I'm not here to test you about astronomy. I know your ignorance of that is astronomical –"

"Do not insult me Dr. Oz."

"OK, then I have a puzzle for you before we enter the next room of our facility. Consider a globe representing Earth." A model of Earth materializes before them and begins to spin so that Dorothy can see various orientations:

Spinning globe

Next, Dr. Oz hands Dorothy a piece of Silly Putty in the shape and size of Kansas on the globe. "Dorothy, you are to toss the putty at the spinning globe. If, during your random toss, your putty Kansas touches Kansas, you may return home. Here is my question. What are the chances of returning home on your first toss? I will be happy if you give me a good estimate."

Difficulty Level: ✳ ✳

4 An Experiment with Signs

The next morning the sun was behind a cloud, but they started on, as if they were quite sure which way they were going. "If we walk far enough," said Dorothy, "I am sure we shall sometime come to some place."

— Dorothy, *The Wonderful Wizard of Oz*

Dorothy is traveling in a sleek hovercraft along Highway 96 from Ness City to Bazine, Kansas. The Walnut River rages to their right as blackbirds cry overhead.

Dorothy looks out the window. Something is wrong. Dead wrong. There are no people. Occasionally a cluster of robots pass Dorothy and Dr. Oz. The robots grin at Dorothy revealing shiny teeth the color of silver. Perhaps she is in Kansas a thousand years in the future.

Dr. Oz slows the hovercraft as he sees a complex collection of signs.

"Dorothy, here is a test for you that no Earthling has ever passed. Which sign or signs should be removed?"

Dorothy stares at the signs as Toto barks at passing hovercrafts. "This is a difficult problem. I have to decide on an acceptable criterion by which to eliminate the sign."

"You had better think quickly, my dear." Dr. Oz's tentacle snakes closer to Toto as the dog freezes in fear.

Can you help Dorothy?

Difficulty Level: 🐛🐛

5 The Logic of Greenness

Many shops stood in the street, and Dorothy saw that everything in them was green. Green candy and green pop corn were offered for sale, as well as green shoes, green hats, and green clothes of all sorts. At one place a man was selling green lemonade, and when the children bought it Dorothy could see that they paid for it with green pennies.

— A description of Emerald City, *The Wonderful Wizard of Oz*

"Please follow me," Dr. Oz says to Dorothy.

He opens a great green door in the Oz testing facility. Once inside, Dorothy passes through a metal detector operated by two squidlike aliens. Dorothy thinks that these must be trochophores of the same species as Dr. Oz. One of the creatures wears a green Armani suit, the pants held up by suspenders. The other alien is huge, with long, pale green tentacles. His several-hundred-pound body resembles the Great Pyramid of Giza.

Blinking green lights are hanging from the green walls. On bookshelves are a number of mathematical puzzles books: the complete collections of Martin Gardner, Henry Dudeney, Sam Loyd, Angela Dunn, and Charles Trigg. One of the aliens appears deep in thought as he peruses Clifford Pickover's *Wonders of Numbers*. There are also other books, written in an alien language that Dorothy has no possibility of deciphering.

"Dorothy, here is your next puzzle." He motions toward a group of creatures. "Standing before you is a race of aliens with varying degrees of greenness. As you can see, some are very green. Some are pale – almost white. In this rectangular array of aliens, which will be greener, the greenest of the palest aliens in each column, or the palest of the greenest aliens in each row? I want an answer that is always true and not just true for this particular array of aliens before you."

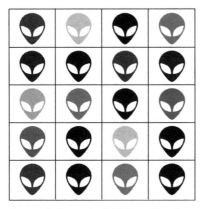

Array of aliens at which Dorothy stares

"Dr. Oz, that's impossible. There can't possibly be enough information to answer your question in general."

"Ah, my pretty, but you are wrong. You do have enough information."

Dorothy tries to imagine different arrays of aliens but can't possibly visualize all the different shades of green.

Can you help her solve the problem?

Difficulty Level: 🔆 🔆 🔆 🔆

6 Magical Maze

"It must be inconvenient to be made of flesh," said the Scarecrow thoughtfully, "for you must sleep, and eat and drink. However, you have brains, and it is worth a lot of bother to be able to think properly."

— The Scarecrow, *The Wonderful Wizard of Oz*

Dr. Oz points to a large plaque on the wall containing a collection of sentences interspersed with little animals symbols. "We call this the Magical Maze. Whoever solves it is said to have a very long life and great wisdom. Alas, no one has been able to solve in under ten minutes. Can you? You start with 10 pebbles in your pocket."

Dorothy steps closer. "What do I do to solve it?"

"The writing on the wall is actually a maze, the object of which is to get from the circular bubble (O) at upper left to the matching bubble on the bottom line. Along the way, there are various animal shapes dividing the text into single sentences, each of which is an instruction for you to follow. As you try to get from the start to the end (left to right, as if reading a book), you must obey each instruction you reach, unless you are instructed not to obey. Good luck!"

(○) Start here. 🐌 Add 5 pebbles to your pocket. 🐦 Jump to any bird. 🦓 If you have a piece of gum in your pocket, jump to a rabbit, otherwise jump to a seahorse. 🐘 Add 11 pebbles to your pocket. 🐢 If this sentence has an even number of words, jump to any seahorse. 🦢 Add a piece of gum to your pocket. 🐎 If you do not have a piece of gum in your pocket, jump to a zebra. 🦢 Add 8 stones to your pocket. 🐒 Add 30 pieces of gum to your pocket. 🦓 Jump to any seahorse. 🐒 Add 12 pebbles and one stick of gum to your pocket. 🐎 Jump to a turtle. 🐈 If you have exactly one piece of gum in your pocket, eat it and jump to a seahorse. 🦌 Add 30 pebbles to your pocket and jump to a dog. 🐐 Add 4 pebbles to your pocket and go to any zebra. 🐎 If you have a piece of gum in your pocket, jump to a zebra; otherwise jump to a turtle. 🐈 Do not subtract pebbles whenever you are told to do so. 🦋 Subtract 5 pebbles and go to the start. 🐐 If you have over 100 pebbles, jump to any butterfly. 🦋 If you are human, jump to any squirrel. 🦋 If you have over exactly 400 pieces of gum, jump to any butterfly. 🐈 Add 13 pebbles to your pocket, one million pieces of gum, and then jump to any zebra. 🦌 If you have over 38 pebbles, go to the next squirrel. 🦋 If you have an odd number of pebbles, go to a butterfly. (○) End here! Congratulations. You did it!

Difficulty Level: 🌸 🌸

7 Kansas Railway Contraction

"I have heard, my dear friend, that a person can become over-educated; and although I have a high respect for brains, no matter how they may be arranged or classified, I begin to suspect that yours are slightly tangled. In any event, I must beg you to restrain your superior education while in our society."

— The Scarecrow, *The Marvelous Land of Oz*

Dr. Oz, Dorothy, and Toto have left the confines of the Oz testing facility and are now riding on the Midland Railway from Baldwin City to Norwood, Kansas. Dorothy is enjoying the scenic Eastern Kansas farmland and woods. She knows *where* she is; she's not sure *when.*

Dr. Oz turns to Dorothy. "Even this mundane, primitive system of transport must have seemed alien once. Yet its production. . . ."

Dorothy nods. "The first improvement on cast-iron rails were rails of wrought iron, introduced in 1820 in England, where the first steel rails were also manufactured."

Dr. Oz takes a step back. "Ooh. How did you know all this?"

"One of your colleagues gave me a pill that increased my brain power." Dorothy pauses. "The manufacture of steel rails in the United States began in 1865, and the rails are now used throughout the world." Dorothy looks directly into Dr. Oz's eyes. "But, of course, you must have known that, Mr. Squid. You're the know-it-all."

"My name is *Dr. Oz,* not 'Mr. Squid.' Yes, I was just testing you. Now I want you to imagine a railway track made of cast iron and stretching from New York City to Los Angeles."

Long track

Dorothy nods. "Yes, I can imagine such a thing."

"Good. As you know, temperature changes cause metals to expand and contract. Now, I want you to estimate the change in the length of railway track from New York City to Los Angeles as result of a temperature change of −20 degrees F to 110 degrees F."

"Dr. Oz, this is not too realistic, since it is never −20 degrees F in Los Angeles."

"Bear with me. I'm trying to simplify the model for you. Given this temperature change along the entire track, what is your guess for the change in overall track length? Do you think we are talking inches, feet, or miles of length change?"

Dorothy thinks about the problem, but before she can respond in any reasonable fashion, Dr. Oz places her leg within the hole of a medieval torture device resembling a doughnut-shaped piece of solid iron.

Ring of iron around Dorothy's leg

"Dr. Oz, what are you doing?"

"Dorothy, I have a second question for you, also related to temperature. If you were to heat the ring of iron around your leg, would that help you escape?"

Dorothy reasons that heating the iron will cause it to expand, but she is not sure if the hole will become larger, smaller, or remain the same size.

Can you help her escape from the ring? Also, what is your guess about the train-track expansion?

Difficulty Level: 🐜🐜

8 The Problem of the Bones

B.C. MANSFIELD

"I consider brains far superior to money, in every way. You may have noticed that if one has money without brains, he cannot use it to advantage; but if one has brains without money, they will enable him to live comfortably to the end of his days."

— The Scarecrow, *The Marvelous Land of Oz*

On either side of the road are fields dotted with thickets of ghost-like trees of amber and ochre. The road becomes narrower and finally ends next to a large rock. Leaning on the rock is a large gray alien with a cadaverous face. His hands look skeletal. His nails are long and ivory white.

Dorothy hides behind Dr. Oz. "Who – who is that?"

"We call him the Bone Being," Dr. Oz says with a shudder. "Look into his pit."

Dorothy and Dr. Oz peer into a deep hole in the ground. The Bone Being comes closer and opens and closes his mouth spasmodically. "In the pit," he says, "are 10,000 leg bones. I have cracked each bone at random into two pieces by throwing it against a rock. What do you think is the average ratio of the length of the long piece to the length of the short piece for each time I crack a bone? You can reason from a purely theoretical standpoint. If you cannot find the solution within two days, I will add your leg bone to the pit" (Figure 8.1).

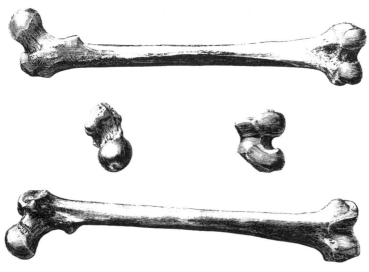

8.1. Human leg bones.

Dorothy tries an experiment. She takes a long bone and smashes it into two pieces. The long piece is 1.5 feet long, and the small piece is 0.3 feet long. The ratio of the length of the long piece to the short piece is therefore 5. But is this typical?

How can you solve this more generally? If you were forced to gamble on a likely ratio of long piece to short piece produced during a bone break, what ratio would you tell the Bone Being?

Difficulty Level: 🐜 🐜 🐜 🐜

9 Square Overdrive

"I suppose I must start my brains working," replied his Majesty the Scarecrow; "for experience has taught me that I can do anything if I but take time to think it out."

— The Scarecrow, *The Marvelous Land of Oz*

Dorothy, Toto, and Dr. Oz are walking through the Oz testing facility. There are lots of odd sounds: an electrical hum; the sloshing of water or other liquids; Toto's clattering nails against a hard floor. Dr. Oz comes to an abrupt stop.

"Dorothy, here are two questions for you today. Take your time. You have all day to answer them."

"Continue," says Dorothy.

"*Problem 1:* What integer number added separately to 100, 200, and 300, will make the three results three different *square numbers? Problem 2:* What integer number added separately to 100, 101, and 102, will make the three results three different *cube numbers?*"

Dorothy looks at Dr. Oz. "I've always been fascinated by square numbers like 0, 1, 4, 9, 25, and 36. They're called 'square numbers' because they result from squaring some integer, for example, $0^2 = 0$, $1^2 = 1$, $2^2 = 4$, $3^2 = 9$, $5^2 = 25$, and $6^2 = 36$. Similarly, cube number are numbers like 8 and 27 because these numbers result from cubing some integer. In these two examples, 2^3 is 8 and 3^3 is 27."

"Oooh." Dr. Oz's tentacles quiver. Apparently he is excited by Dorothy's mental agility. However, he quickly recovers his composure.

"Okay, then. Because you know so much, I will add a third question to today's set."

"That's not fair!"

"Fair? I'll show you fair." Dr. Oz uses his anterior tentacle to lift Toto into the air.

"Wait," Dorothy says as she assumes a defensive Kung Fu "crane stance" that Aunt Em had taught her. Dorothy's hands become as straight and stiff as knives. "There's no need to become violent. I will undertake your third problem."

Dr. Oz places Toto gently on the floor. "One question that I've pondered is whether or not it is possible to fill an infinite square array with distinct integers such that the sum of the squares of any two adjacent numbers is also a square. To illustrate, here is a 4 × 4 array with the desired property." Dr. Oz paints the following array on the floor using ink secreted from one of his posterior orifices.

1836	105	252	735
1248	100	240	700
936	75	180	525
273	560	1344	3920

Largest "squarion" array?

"For example $75^2 + 180^2 = 195^2$. Is it possible to create bigger arrays of this kind? Can you?"

10 Squares and Cubes

"Pardon me," returned the Scarecrow. "My brains are slightly mixed since I was last laundered. Would it be improper for me to ask, also, what the 'T.E.' at the end of your name stands for?"

"Those letters express my degree," answered the Woggle-Bug, with a condescending smile. "To be more explicit, the initials mean that I am Thoroughly Educated."

— The Marvelous Land of Oz

Dorothy and Dr. Oz are walking along the shores of Milford Lake, one of Kansas's largest bodies of water. Dr. Oz leaps into the lake to moisturize his rubbery skin. When he returns, he seems refreshed and happy. "Dorothy, my previous test question involved square and cube numbers, which you seem to be interested in. I have another problem involving such numbers."

Dorothy tosses a stone into the water. "Please continue."

"Find three different integers such that the sum of their squares equals a cube number, and the sum of their cubes equals a square."

As Dorothy gazes into the lake, she begins to try out different combinations of numbers. For example, she tries 1, 2, and 3. The sum of their squares is $1^2 + 2^2 + 3^2 = 14$. Unfortunately, 14 is not a cube number, so 1, 2, and 3 don't even satisfy the first part of Dr. Oz's conditions that the sum of the squares be a cube number. This certainly seems like a difficult problem for Dorothy. Is there a solution?

Difficulty Level: 🐜 🐜

11 Mr. Plex's Matrix

"I do not blame you," said the Scarecrow. "Education is a thing to be proud of. I'm educated myself. The mess of brains given me by the Great Wizard is considered by my friends to be unexcelled."

— The Marvelous Land of Oz

Dr. Oz and Dorothy are riding a hovercraft though Chase County, one of Kansas's largest regions of unspoiled prairie. Some of the hills look like they are all rock without enough soil to support life, but Dorothy knows that in the springtime this desolate view will be replaced by a thick carpet of wildflowers and lush bluestem grasses.

Dr. Oz parks his hovercraft and hands Dorothy a card of symbols:

Dr. Oz's symbol card

Dorothy takes the card in her hand. "Are those symbols letters used by some exotic, intelligent alien race?"

"I doubt it. This is a puzzle for you. Are you ready?"

"Go ahead."

"What symbol should be used to replace the question mark in the matrix of symbols? Here's a hint for you: numerical values need to be assigned to the symbols to solve this."

"Dr. Oz, who's that odd creature with the large, toothy grin." She points to the symbol:

"Oh, that's my cousin, Mr. Plex. He gets on my nerves sometimes. He is always trying to insert himself into my puzzles. A real jokester, that Mr. Plex. Now, can you solve the puzzle? The symbol for Mr. Plex should be treated just like any of the other symbols. That means he is also assigned a numerical value."

Dorothy sits on the hovercraft's hood and begins to ponder the bizarre array. "This is too hard."

"OK, here's a hint. I've filled in a few of the cells for you. To solve the problem you must discover a common mathematical rule that applies to each row."

Can you help Dorothy? What is the logic you used to solve this puzzle? Is there another logic that you might use to solve it differently?

4	3	ß	F	ß
🌀	1	ß	F	🌀
🌀	ß	⋈	F	⋈
ß	ß	ß	F	?
F	F	⋈	ß	ß

Dr. Oz's help

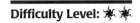

12 Chaos in the Clock Factory

If a lunatic scribbles a jumble of mathematical symbols it does not follow that the writing means anything merely because to the inexpert eye it is indistinguishable from higher mathematics.

— Eric Temple Bell, in J. R. Newman's *The World of Mathematics*

Dr. Oz and Dorothy are visiting an eccentric employee of a clock company who has arranged clocks in the manner shown in Figure 12.1.

Dorothy looks at the attractive arrangement of clocks. "Why is the clockmaker behaving so strangely?"

"We are experimenting with his brain. Don't worry, we will return him to his former self in a few hours. But Dorothy, we're not here to discuss the lunatic clockmaker. Look at his pattern of clocks. Little does the man realize that the arrangement of clocks creates an interesting maze. Starting at the clock at the upper right, you may travel to the goal at the bottom in the following way."

"Wait, slow down."

Dr. Oz nods. "To solve the maze, you move from one clock face to an adjacent clock. Adjacent clocks share one edge. If an adjacent clock face has an arrow opposing your movement, then you may not move into the adjacent clock, because you can't travel against the arrow's direction."

"Dr. Oz, what do you mean by oppose?"

Dr. Oz draws two clocks. "You can't travel from either clock to the

12.1. Travel from Start to End using the rules described. (Illustration by Brian Mansfield.)

other – from the left to right, or right to left – because the destination clock's hour hand *opposes* you, it points the opposite way."

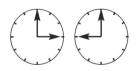

Can't travel horizontally

Dr. Oz draws another set of clocks.

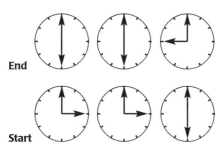

End

Start

"Here, you could start at the clock at the lower left, and then proceed to the right, and then up, and finally go left to the finish clock. At no point in the path did you have to go against a clock hand. Assume that all hands have a pointy arrow at their ends, and you don't want to run into the sharp arrowhead."

Dr. Oz takes a digital snapshot of a small segment of the maze and labels four clocks A, B, C, and D (Figure 12.2). "Starting from the clock marked A you can travel to clock B or clock C, but not D, because D's hand opposes A's hand. Now, here is your challenge. What's the *longest* path you can find to the goal? You can't cross through a clock more than once. (For a truly difficult brain boggler, is there a way to reach the goal if all clock hands are rotated by 90 degrees? How about 180 degrees, or 270 degrees?)"

12.2. Close-up of a region of 12.1.

13 The Upsilon Configuration

"No one really understood music unless he was a scientist," her father had declared, and not just a scientist, either, oh, no, only the real ones, the theoreticians, whose language is mathematics. She had not understood mathematics until he had explained to her that it was the symbolic language of relationships. "And relationships," he had told her, "contained the essential meaning of life."

— Pearl S. Buck, *The Goddess Abides*

Dorothy is munching on several of Toto's nutritious dog biscuits because Dr. Oz will not give her ample human food until she solves a puzzle involving steps (Figure 13.1).

"Dorothy, put down your food," Dr. Oz says. "This difficult set of steps and cliffs is called the Upsilon Configuration after a trochophore named Upsilon who was able to solve it within one hour. I will give you a delicious filet mignon steak or sushi, your choice, if you can follow my exact instructions. You must step on as many blocks as possible in the Upsilon Configuration without stepping on any block more than once. You may step up or down one level at a time (as long as the two blocks you step from and to share an edge) or to a neighboring block on the same level. I don't allow you to move diagonally."

Dorothy drops her biscuits and nods. "OK, I will step on every step I can. But I must say that you are a cruel taskmaster."

"Wait! There are two extra conditions. You must also pick up the clip, helmet, and rope in this order! And the blocks are slippery with oil. One false move and you're dead."

13.1. Upsilon Configuration. (Illustration by Brian Mansfield.)

"Oh, you lout!"

"Remember, you must cover the maximum number of blocks possible using my rules!"

Difficulty Level: 🐜🐜

14 Bone Toss

B.C. MANSFIELD

"Everything in life is unusual until you get accustomed to it," answered the Scarecrow.

"What rare philosophy!" exclaimed the Woggle-Bug, admiringly.

"Yes; my brains are working well today," admitted the Scarecrow, an accent of pride in his voice.

— The Marvelous Land of Oz

Dorothy, Toto, and Dr. Oz are outside the testing facility and walking along a dirt road. Trees are becoming more common, and soon forests beside the road are dark with mystery. Dorothy sees clear streams, and Dr. Oz permits Dorothy to stop and get a drink.

Suddenly, an alien comes up to her and draws a circle of radius R on the ground, while handing a thin leg bone to Dorothy. "I'd like you to toss a bone of length R at the circle," the ambidextrous amphibian says, "so that one end of the bone is on the circle's edge, but the bone's orientation is completely random."

"I'm not touching that disgusting thing," Dorothy says.

"OK," the alien says, "try to stay calm. You can use a stick instead." He hands the bone to Toto, who takes it aside and begins to gnaw.

Dorothy tosses a stick and finally one end of the stick happens to touch the edge of the circle. In this case, the other end of the stick also happens to be inside the circle.

"Very good," says Dr. Oz. "Here is your test. In general, what is the probability that once one end of the stick is on the circle that the other end of the stick is *inside* the circle? How would your answer change if the thin stick were of length $2R$ or $R/2$? If you do not respond properly, the alien may toss you into the circle and keep you prisoner."

Difficulty Level: ✹ ✹ ✹

15 Animal Farm Courthouse

"At the same time," declared the Tin Woodman, "you must acknowledge that a good heart is a thing that brains can not create, and that money can not buy. Perhaps, after all, it is I who am the richest man in all the world."

— The Marvelous Land of Oz

Dorothy and Dr. Oz are walking through the courthouse at the end of Broadway, the main business street in Cottonwood Falls, Kansas. It's the oldest courthouse west of the Mississippi. They walk past the wooden staircase and the jail. In the distance Dorothy sees an ornate bridge over the Cottonwood River.

There's just one problem. The aliens have evacuated Cottonwood Falls. The courthouse has been turned into a makeshift zoo that permits the aliens to study an array of Earthly animals.

Dr. Oz points to a viewscreen that the aliens must have hastily hung along the wooden staircase. "Dorothy, my colleagues in the courthouse wish me to give you a mazelike problem. You are to find the longest path possible through the zoo depicted on the viewscreen."

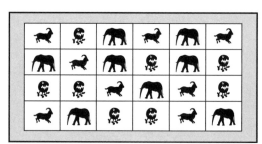

Zoo maze

"How am I supposed to travel through the maze?"

"You may start anywhere, and each move must be either up, down, right, or left. Your path must repeat any three-animal pattern, such as

🐐 🐘 🦔, 🐐 🐘 🦔, etc.,

for as long as possible, without ever crossing itself.

"What is the toothy Mr. Plex doing in your zoo?"

"We've temporarily imprisoned Mr. Plex and a few of his siblings in the zoo. We're trying to teach him a lesson so he stops invading my puzzles. We'll release him if you can solve this puzzle."

Difficulty Level: 🕷 🕷

16 The Omega Sphere

"That is hard to tell," said the man thoughtfully. "You see, Oz is a Great Wizard, and can take on any form he wishes. So that some say he looks like a bird; and some say he looks like an elephant; and some say he looks like a cat. To others he appears as a beautiful fairy, or a brownie, or in any other form that pleases him. But who the real Oz is, when he is in his own form, no living person can tell."

— The Marvelous Land of Oz

Dr. Oz walks up to Dorothy and hands her a glowing sphere about the size of a basketball (Figure 16.1).

"What's that?" she says, as Toto barks at Dr. Oz.

He extends his tentacle. "Inside the sphere is a random collection of two billion points. Does there exist a plane having exactly one billion of these points on each side of the plane? If so, why?"

Difficulty Level: ✸ ✸

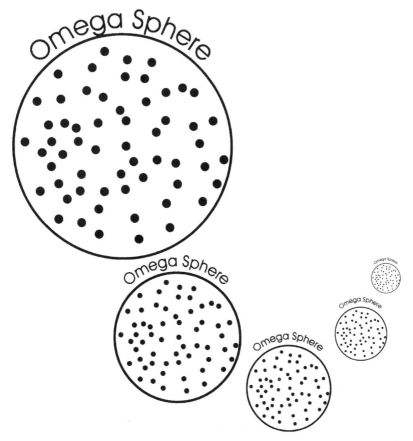

16.1. Several Omega Spheres. (Each holds two billion points with random positions.)

17 Leg-Bone Shatter Produces Triangle

Intellectuals solve problems; geniuses prevent them.

— Albert Einstein

Dorothy, Toto, and Dr. Oz are wading through a swamp when they suddenly come upon the Bone Being – the cadaverous creature whom they had encountered earlier (Chapter 8).

"Hello," he says using his crude organs of speech. He carefully moves the sharp edge of his claw across the chest of Toto.

"Keep away from him, Bone Being!"

The Bone Being wipes some of Toto's hair off his claw with a towel and draws back the corners of his mouth in laughter, revealing fangs.

"I have another problem for you. Follow me." He is utterly white and smooth. His head and abdomen looks as if he is sculpted from bleached bones.

They walk upon dry ground as the Bone Being grabs a leg bone that he has placed behind a tree. "Think of the bone as being N feet long," he says, and then he shatters the leg into three pieces at random.

"Dorothy, what is the probability that these three pieces can be placed together to form a triangle?"

"That is too difficult for her," says Dr. Oz.

Dorothy puts her hands on her hips. "No it isn't."

The Bone Being nods. "For the second part of this problem, I give you the longest piece of a recent shatter experiment. If you were a betting person, what is the most likely ratio of the longest piece of bone to the shortest piece of bone in each shattered set? How did you solve this?"

18 Z-Bar Ranch

One definition of man is "an intelligence served by organs."

— Ralph Waldo Emerson

Dorothy and Dr. Oz are strolling through the Z-Bar Ranch near Strong City, Kansas. It's located in the Tallgrass Prairie National Preserve and was built in 1881.

Within minutes, Dorothy enters a large stone house surrounded by 11,000 never-plowed acres of prairie that form the heart of the prairie park.

"Dr. Oz, why are you taking me here?"

"This is one of my secret outposts. Mr. Plex is will serve as my assistant. He has gathered several animals and a clone of himself together for a puzzle. Look over there."

Dorothy looks out and sees a 6-by-6 pen.

"Dorothy, imagine that all the creatures are lined up on a piece of paper that you can cut with a scissors. Your task is to create two areas, both of exactly the same size and shape, so that both areas contain equal numbers of each kind of creature."

19   The Mystery of Phasers

B.C. MANSFIELD

Intelligence is that faculty of mind, by which order is perceived in a situation previously considered disordered.

— Haneef Fatmi, "A Definition of Intelligence," *Nature*

Dorothy is with Dr. Oz in a huge starship shaped like a teardrop. They are racing toward the ship of the cadaverous Bone Being (mentioned in previous chapters), which looks like a mile-long skull.

"Dr. Oz, this is dangerous!" Dorothy screams. "What have you gotten me into now?" Toto hides in the corner beneath a robot that resembles a hybrid of Britney Spears and Christina Aguilera.

Dorothy extends his tentacle toward Dorothy. "This is my latest puzzle to assess your species' intelligence. We are now traveling toward another ship, and both of us are firing our phasers."

"What's a phaser?"

"Oh, you *are* a child. A phaser is a weapon. It's not one of those bogus ray guns you might have seen in *Star Trek.* We use a specific UV frequency of 193 nanometers establishing an invisible wire of light through the air, from the transmitter to the target, wherein an electrical current could be imparted."

Dorothy rolls her eyes. "Oh, *that* certainly clears things up. As if I should know what 'nanometer' means?"

"Never mind that. Just listen to the problem. Each time a ship fires it either misses the opponent ship or destroys it totally. Also, each time a ship fires, it has a 50% chance of hitting the opponent ship. For the sake of discussion, call our ship 'ship *A*' and the Bone Being's ship 'ship *B*.' First ship *A* fires, then ship *B*, then ship *A*, alternating until one ship is destroyed. What are the odds that ship *A* will survive? How do these odds change if each time a ship fires, it has only 10% chance of hitting the opponent ship?"

"I don't like this scenario."

Dr. Oz nods. "Now here's something even more difficult. I call it the Dreaded Fibonacci Gambit." Dr. Oz hands Dorothy a card with the words:

Dr. Oz continues. "In this DFG question, the ships fire at one another according to the Fibonacci sequence: 1, 1, 2, 3, 5, 8, 13, 21, 34, 55, 89, 144, 233, 377. (Starting with "2" each successive number is the sum of the two previous ones.) In other words, first ship *A* fires 1 shot, ship *B* fires 1 shot, ship *A* fires 2 shots, and so forth. What are the odds that ship *A* will survive for the 50% and 10% scenario? How did you solve this?"

"Oh God, too many questions!"

"You have three problems to solve. If you get all three correct, I'll take you back to Earth."

Difficulty Level: ✺ ✺ ✺ ✺

20 Salty Number Cycle

Intellectual brilliance is no guarantee against being dead wrong.

— David Fasold

"What's that?" Dorothy says as she points to a large mound of white material.

Dr. Oz runs his tentacles lingeringly over the powdery substance. "The world's largest salt deposit is here – Hutchinson, Kansas. This deposit is 100 miles by 40 miles, 325 feet thick, and yields 44.1 million tons of salt each year. We have a mathematical testing facility 650 feet underground. Nature maintains a constant temperature of 68 degrees with virtually no humidity. This environment is ideally suited for our record storage."

As they walk, Dorothy and Dr. Oz pass various trochophores with calculators in their tentacles. Mr. Plex emerges from a dark tunnel.

"Sir, I have a new puzzle for Dorothy." Mr. Plex hands a card to Dorothy and Dr. Oz.

Number cycle

Dr. Oz points to its center. "Mr. Plex, must you insert your beautiful physiognomy in every problem?"

"Sir, never mind that. Going clockwise or counterclockwise, the ring of numbers forms a mathematical expression when certain empty cells are filled with +, −, ÷, ×, =, ↑ (exponentiation or raising to a power), and E (end of expression). How many solutions, if any, can you find? You can leave cells blank, but if you do not place a symbol between two numbers, they form a multidigit number. For example, 2 1 would become 21."

Difficulty Level: 🌟

21 Where Are the Composites?

We arrive at truth, not by reason only, but also by the heart.

— Blaise Pascal, *Pensées* (1670)

Dr. Oz is playing with Toto and feeding him squid-shaped dog biscuits. "Dorothy, today I want you to find 10,000 consecutive non-prime numbers. When you find them, tell me how you solved this."

Dorothy grabs Toto and backs away. "That's a tall order. Don't I need a computer to solve this puzzle?"

"No computer is needed if you just give me a very simple recipe for finding the 10,000 numbers. And need I remind you that a prime is a positive integer that cannot be written as the product of two or more smaller integers? For example, the number 6 is equal to 2 times 3; therefore, it is not prime. 6 is a nonprime or "composite" number. On the other hand, 7 cannot be written as a product of two or smaller integers; therefore, 7 is called a prime number or prime."

"Dr. Oz, you are quite talkative today."

He nods. "Here are the first few prime numbers: 2, 3, 5, 7, 11, 13, 17, 19, 23, 29, 31, 37, 41, 43, 47, 53, 59. Notice that the gaps between successive prime numbers varies; for example, between these first few primes, the gaps are 1, 2, 2, 4, 2, 4, 2, 4, 6, 2, 6, 4, 2, 4, 6, 6. . . . The Greek mathematician Euclid proved that there is an infinite number of prime numbers. But these numbers do not occur in a regular sequence, and there is no formula for generating them. Therefore, the discovery of large new primes requires generating and testing millions of numbers."

Dr. Oz tosses a dog biscuit into his moist gullet. "With this little introduction to prime and nonprime numbers, your mission is to find 10,000 consecutive nonprime numbers. How do you solve this?"

Difficulty Level: ✹ ✹ ✹

22 Brain Trip

The brain is a three-pound mass you can hold in your hand that can conceive of a universe a hundred-billion light-years across.

— Marian Diamond, OMNI

Dorothy and Dr. Oz are in a tiny submarine inside a complicated electronic brain that straddles the Kansas–Oklahoma border. Dorothy wants to escape from the brain. The brain's defense system does not know that Dorothy is there to effect repairs and be helpful, and therefore the brain wants to destroy her. The brain has placed many charging and decharging stations (symbolized by + and –) along her possible escape routes (Figure 22.1).

"Dr. Oz, this is nuts!"

"I know. That's why it is a good challenge for you."

Dorothy travels through the minefield. . . . If she goes through two charging stations in succession (+, +), her craft will receive too much electricity and burn out. Similarly, if she goes through two decharging stations in succession (–, –), her reserves will be drained and her craft stop functioning. Therefore, she must travel in such a way as to alternate between decharging and charging stations. (Incidentally, she can't make a U-turn once she touches a charging station).

Can she escape from the electronic brain?

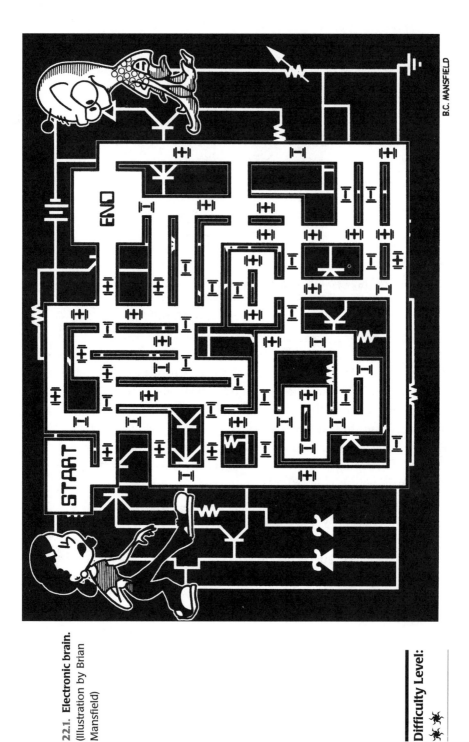

22.1. Electronic brain.
(Illustration by Brian Mansfield)

B.C. MANSFIELD

Difficulty Level:

23 The Gaps of Omicron

The difference between a smart man and a wise man is that a smart man knows what to say, and a wise man knows whether or not to say it.

— Frank M. Garafola

Dr. Oz is walking with Dorothy in an abandoned schoolhouse. "Dorothy, did you know that sugar is the only word in English where the letters 'su' are pronounced like 'sh'?"

Dorothy thinks for a moment. "Are you sure?"

Dr. Oz waves his tentacle. "I'll think about it. In the meantime, let's talk mathematics. Today's problem deals with rational numbers."

"Rational numbers?"

"A *rational number* is a number that can be expressed as a ratio of two integers." [*Author's note:* The Further Exploring section for Chapter 9 has more information and examples of rational numbers.]

Dr. Oz hands Dorothy a card with an interesting-looking formula:

$$\alpha^\beta = \beta^\alpha$$

"What can you tell me about the rational solutions to this formula? Can you draw a plot of unequal rational numbers α and β that satisfy the relationship? What pretty patterns, if any, do you see in the plot? How did you solve this?

Difficulty Level: ❄ ❄ ❄ ❄

24 Hutchinson Problem

The intelligent man finds almost everything ridiculous, the sensible man hardly anything.

> — Johann Wolfgang Von Goethe, in "1,911 Best Things Anybody Ever Said," Robert Byrne, editor

Hutchison, Kansas, is the wheat capital of the world. As Dr. Oz and Dorothy approach the city, they see large grain elevators next to railroad tracks. The railroad track seems to run right through the middle of the town.

Dorothy hears a train whistle blowing as a train crosses the roads of Hutchinson. Evidently the alien trochophores have taken over portions of the town, as evidenced by cryptic signs in a tongue that Dorothy can never decipher:

Dr. Oz turns towards Dorothy and hands her a card. "Never mind the signs. I have a new problem for you. Consider Mr. Plex, who must get from the cell at upper left to the empty cell at the lower right by jumping the number of cells indicated by the cell it lands on. Can you find a way? Mr. Plex can move up, down, right, or left. For example, he might start by hopping 2 to the right and then 3 to the right or 3 down. What is the fewest number of jumps required?"

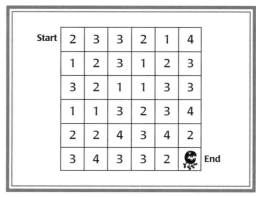

Start	2	3	3	2	1	4	
	1	2	3	1	2	3	
	3	2	1	1	3	3	
	1	1	3	2	3	4	
	2	2	4	3	4	2	
	3	4	3	3	2		End

Hutchinson problem

Difficulty Level: ✷ ✷

25 Flint Hills Series

It is not clear that intelligence has any long-term survival value.

— Stephen Hawking

Dr. Oz and Dorothy are walking through the Flint Hills of Kansas. Occasionally Dr. Oz stumbles. Apparently his deliquescent body is not designed for rocky terrain. "Dorothy, today I want to talk more about series. The word *series* in mathematics usually refers to the sum of a finite or infinite sequence of terms. I like to think of a series from a mountain climber's perspective." Dr. Oz sketches a mountain in the soil:

Mountain

"Mountains? We have no mountains in Kansas. Can you use another analogy?"

"No. Use your imagination. Like a mathematical series, some mountains may eventually level off to a grassy plateau, while others shoot up into the clouds beyond our vision. This analogy should become clearer as I explain further. An infinite series, for example, can be written in the following form." Dr. Oz sketches on the ground:

$$a_1 + a_2 + a_3 + \ldots + a_n + \ldots$$

"This in turn, can be written more compactly."

$$\sum a_n$$

"Here are two examples of infinite series."

$1 - 2 + 3 - 4 + \ldots$ and $1 - \frac{1}{2} + \frac{1}{3} - \frac{1}{4} + \frac{1}{5} \ldots$

Dr. Oz points to the second of the two examples. "The second expression is an example of an oscillating convergent series since the terms are alternately greater and less than the limiting value 0.69314. It's also an 'alternating series' because the signs of the terms alternate. The fact that this series has this limiting value, toward which it tends, implies *convergence*."

Dr. Oz hands Dorothy a card with a computer program on it, and beneath it is the figure of the graph of $1 - \frac{1}{2} + \frac{1}{3} - \frac{1}{4} + \frac{1}{5} \ldots$ as more terms are added. (The graph is shown in Figure 25.1.)

ALGORITHM: How to Compute Oscillating Sign Series
(a.k.a. Alternating Harmonic Series)
```
s = 0
DO i = 1 to 60
    if ((i mod 2) = 0) then t = 1; else t = 1
    v = (t / i)
    s = s + v
    PrintValueFor (i, s)
END
```

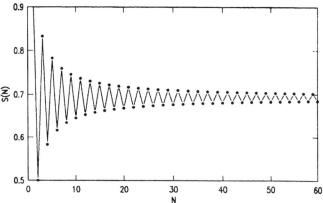

25.1. An oscillating convergent series.

Dorothy looks at the graph. "That's a very zigzagging curve."

"Yes, but many convergent series are less wobbly; that is, they converge in a nonoscillating fashion where the values simply rise (or fall) monotonically to a limit. Think of this limit as the mountain climber's plateau or valley." Dr. Oz pauses. "In contrast to this convergent series, $1 - 2 + 3 - 4 + \dots$ is an example of a *divergent* series because it does not tend to some finite value.

"Often a graph can be used to show the value to which a series converges. Look at the zigzagging graph on my card. The values gradually converge to 0.69314. However, one must be wary of this graphical approach, as evidenced by the fascinating series in an email I intercepted from Ross McPhedran of the Department of Theoretical Physics at Australia's University of Sydney. Consider the infinite series on this card." Dr. Oz hands here the following card:

$$S(N) = \sum_{n=1}^{N} \frac{1}{n^3 \sin^2 n}$$

Flint Hills series

"The Σ symbol indicates summation. Here's an example." Dr. Oz draws in the sand:

$$\sum_{n=1}^{4} n = 1 + 2 + 3 + 4$$

"Dorothy, here is your question. Does the 'Flint Hills series' converge or diverge? You can use any method you like to arrive at your answer."

Difficulty Level: ✹ ✹ ✹ ✹

26 Wacky Tiles

Man is an intelligence, not served by, but in servitude to his organs.

— Aldous Huxley

Dr. Oz is with Dorothy, rowing a boat on Elk City Lake in southeast Kansas. He points toward the water. "Look, lake chickens."

"Canada geese," Dorothy retorts. She watches as a few of the geese take flight when the shadow of alien spacecraft darkens the sky above them. Then she looks more closely at the lake. "I fished here," she says, "before you came to our world."

Dr. Oz nods. "What do all those geese eat?" he says.

"They're grazing animals. They eat grass just like cows. You do have cows on your world?"

"Cows?" Dr. Oz says with a serious expression. "Do you mean those little animals with long white ears?"

"I'm afraid you really don't know much about Earth."

Dr. Oz slams his tentacle down on the boat. "I'll tell you something. I have a puzzle for you." He whirls around and reveals an array of tiles etched on his back. "Fill in the empty tile with the correct two missing symbols."

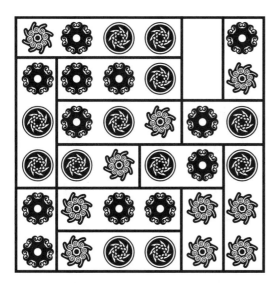

27 Toto Clone Puzzle

My great religion is a belief in the blood, the flesh, as being wiser than the intellect. We can go wrong in our minds. But what our blood feels and believes and says, is always true. The intellect is only a bit and a bridle.

— D. H. Lawrence

Dorothy is standing next to Dr. Oz, who gazes through binoculars at four clones of Toto, each of which stands upon the meadow.

Toto clones prior to their secret arrangement

Dorothy cannot see the Toto clones, but Dr. Oz tells her that the clones are all the same distance away from each other. In other words, the distance between any two of the four Totos is the same. How can this be so?

Dorothy imagines a square array of Totos, as shown below, but they don't satisfy Dr. Oz's requirements because the distance between any two clones is not the same.

Toto test pattern

Dorothy has ten minutes to answer, or Dr. Oz will remove her hair ribbon, never to return it again. What is the arrangement of Totos? If Dorothy answers correctly, she will be set free.

Difficulty Level: 🕷 🕷

28 **Legion's Number**

People who are smart get into Mensa. People who are really smart look around and leave.

— James Randi

"**D**r. Oz, I've solved some of your problems. Why aren't you setting me free?"

"You'll have to solve a lot more than you have, my pretty."

"That's not fair. You're not living up to your promises."

"Never mind that. You get to go free eventually." Dr. Oz hands Dorothy a card. "Dorothy, today your task is to tell me the last 10 digits of the number on the card."

$$N = 666^{666}$$

Legion's number of the first kind

"We call this outrageously large number Legion's number of the first kind. Our philosophers try to know as much as they can about these numbers. They may hold the keys to all the mysteries of the universe."

"Why do you call this 'Legion's number?'"

"Legion was a devil in your New Testament. (Though we religiously pursue interesting puzzles, interesting religions are another pleasingly puzzling pursuit.) In Mark 5, Jesus encounters a multitude of demons possessing a man. When Jesus asks them, 'What is your name?' the man replies, 'My name is Legion, for we are many.' Legion's number is so large it scares me sometime."

"Dr. Oz, you said 'first kind.' That implies there is a second kind of Legion's number."

Legion's number of the second kind

"Yes, your second task it to tell me the last 10 digits of the following number."

"Dorothy, as you know the exclamation point is the factorial sign: $n! = 1 \times 2 \times 3 \times 4 \times \ldots \times n$, so, for example, $3! = 1 \times 2 \times 3 = 6$. These numbers grow fast. For example, $14! = 87,178,291,200$."

"Dr. Oz. Are you nuts? These numbers are huge. I can't calculate them."

"Maybe not, but you can answer my question."

Difficulty Level: ✶ ✶ ✶ ✶

29 The Problem of the Tombs

The test of a first-rate intelligence is the ability to hold two opposed ideas at the same time, and still retain the ability to function.

— F. Scott Fitzgerald

"Follow me," Dr. Oz says as he leads Dorothy and Toto through some dark tombs. "Let's explore."

Dorothy shines a light into the opening between connected tombs "It looks like a cave. It's beautiful." The surface of the tomb walls are tan, glittering with gypsum crystals. The air smells clean and wet, like hair after it is freshly shampooed.

Huddled together like little hobbits are smaller stalagmites of calcite. The larger ones look like the ribs of some giant prehistoric creature.

Dr. Oz shines the flashlight all around them. "This is incredible." Glittering blue gypsum chandeliers, at least 25 feet in length, are suspended above their heads. Walls encrusted with fragile violet aragonite "bushes" line their path. With just a few steps, they've entered another world.

Dr. Oz pauses and looks at Dorothy. "Here is your next puzzle. We are in ten interconnected tombs, $A, B, C, D, E, F, G, H, I, J$. A single short tunnel connects tomb A to B, another B to $C, \ldots$, and a final one I to J. The area of the floor of each tomb corresponds to the Fibonacci sequence, so that the tomb A has a floor with 1 square mile area, tomb B has a floor space of 1 square mile, tomb C has a floor space of 2 square miles, and so forth up to tomb J.

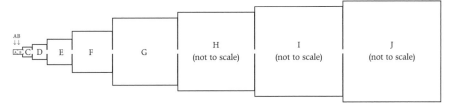

Tombs (schematic)

"Assume that I will introduce 10,000 people into tomb *A,* and they randomly wander the tombs. After a long time, where do you expect to find the most people?"

Dorothy waves her flashlight around. "Shall I assume that the humans randomly move through the tombs, and that there is no escape from the tombs?"

"Yes. For the second part of the problem, consider the same problem as I just explained to you. However, for this second part, assume that the humans behave like actual humans, whose wandering will not be random. Where do you expect to find the most humans after a long time? In this second problem assume that food in the form of 'manna' – a fine flaky substance, as fine as frost, I understand – is distributed uniformly throughout the tombs and that it is in continual supply. Also assume that there is a continual supply of water in the form of fountains uniformly distributed throughout the tombs. Further assume that there are 30 feet of dirt on the bottom of each floor, permitting burial of wastes and dead. The humans will be in here for years."

"Yuck," says Dorothy.

"The dirt contains the usual decay-producing organisms. Where do you find the most humans after a long time? How would your answer differ if (a) each tomb has a square cross section and (b) each tomb has a regular polygonal cross section, such that tomb *A* has a floor in the shape of an equilateral triangle, *B* has a square floor, *C* has a regular pentagonal floor, and so forth?"

Difficulty Level: ✻ ✻ ✻

The Problem of the Tombs   67

30 Mr. Plex's Tiles

The eye sees only what the mind is prepared to comprehend.

— Henri L. Bergson

Mr. Plex and his clones are playing with some alien snakes. "Dorothy, would you like to hold a snake?"

"Ugh, no thanks. They look dangerous."

"Not really. Dr. Oz had me set up a room with glass walls so that you can see all my clones and snakes. Think of this as a set of tiles."

Mr. Plex points to a viewscreen that shows an aerial view of the snakes and clones.

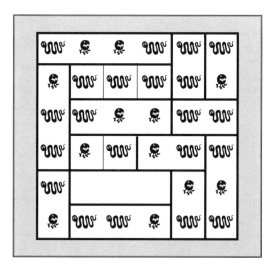

"Your job is to fill in the empty tile with the correct three missing aliens. If you can solve the problem within ten minutes, Dr. Oz will permanently clone your dog Toto so that you can have two companions instead of one. If you do not solve the problem, Dr. Oz will unleash thousands of alien snakes in prairies of Kansas. There's no telling what ecological damage that would do."

Difficulty Level: 🐛🐛

31 Phasers on Targets

B.C. MANSFIELD

The empires of the futures are the empires of the mind.

– Winston Churchill

Dr. Oz and Dorothy are at a shooting gallery in the Oz testing facility.

"Dr. Oz, thank you for letting me try your phaser."

"You're welcome. Just don't try to aim it at me and try to escape, or my colleague Mr. Plex will be very angry with you and hunt you down."

"Why do you have to be so mean?"

"Don't worry. When we're all done with these mathematical puzzles, I'll set you free. Now, I want you to fire at the green target hanging on the wall."

Dorothy aims the phaser at the center of a circular target and fires once. She pauses and fires again. Darn, her second shot is even further from the center than the first.

Target

"Dorothy, you have one more shot. Before you take it, tell me what is the probability that your final shot will be further from the center than your first? Assume that your skill level stays constant."

"Dr. Oz, how can I give you an answer? I don't seem to have enough information."

"You do have enough information. For part 2 of this problem, assume you fire 1000 times. What is the probability your final shot will be further from the center than your first? Are you confident of placing a $100 wager on this? Would you wager Toto? Is your answer different if the target is an equilateral triangle? Does it matter that only you are playing the game, or would your answer be the same if I alternated shots with you?"

Difficulty Level: 🐞 🐞

32 The Chamber of Death and Despair

"How are your brains?" inquired the little humbug, as he grasped the soft, stuffed hands of his old friend.

"Working finely," answered the Scarecrow. "I'm very certain, Oz, that you gave me the best brains in the world, for I can think with them day and night, when all other brains are fast asleep."

— Dorothy and the Wizard in Oz

Dr. Oz places Dorothy and Mr. Plex in a subterranean maze. Dorothy lets her gaze drift to the pockets of crystals around them, noting the flowing harmony of the fractal formations and the crystalline outcropping of rock coated with strips of velvet purple. A cool peace floods her, and for a second the memory of her abduction by Dr. Oz seems to ebb away.

"There are three exit tunnels," Dr. Oz says, "and at the end of each are three Toto clones." He hands Dorothy a card showing the layout of the tunnels."

Dorothy is at the position marked "Start" and cannot actually see the three Toto clones from where she now stands, but she listens intently to Dr. Oz's speech.

Dr. Oz continues. "Two of the three tunnels are coated with a poison and if you travel them you will eventually become trapped in a sticky substance and die. One of the tunnels leads to freedom."

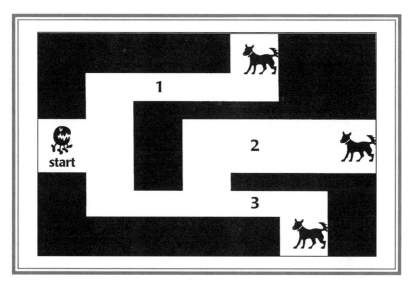

Chamber of death and despair

Dorothy thinks about it for a few seconds, and with fear and trepidation says, "OK, I have no reason for my answer, but I choose Tunnel 1. It's as good as any."

Dr. Oz, who knows where each tunnel leads, extends his snaky tentacle and slowly points to Tunnel 3 and truthfully tells Dorothy that Tunnel 3 is a sticky tunnel.

"Dorothy, I'll give you the opportunity of changing your mind and choosing Tunnel 2 instead of your first choice, if you wish. You can assume that, whenever I test humans, I always point to a tunnel that is a sticky route. Now, here is your question. Is it to your advantage to switch your choice of escape route from Tunnel 1 to Tunnel 2?"

Difficulty Level: ✴ ✴ ✴

33 Zebra Irrationals

I must tell you that I was born in Omaha, and my father, who was a politician, named me Oscar Zoroaster Phadrig Isaac Norman Henkle Emmannuel Ambroise Diggs, Diggs being the last name because he could think of no more to go before it. Taken altogether, it was a dreadfully long name to weigh down a poor innocent child, and one of the hardest lessons I ever learned was to remember my own name. When I grew up I just called myself O. Z., because the other initials were P-I-N-H-E-A-D; and that spelled 'pinhead,' which was a reflection on my intelligence."

— The Wizard in *Dorothy and the Wizard in Oz*

Dr. Oz, Mr. Plex, and Dorothy are looking at zebras in a zoo. "Dorothy, those strange animals with stripes remind me of a mathematical puzzle."

Zebras

"Dr. Oz, we call them 'zebras.'"

"The ancient Greeks discovered that some naturally occurring lengths, such as the diagonal of a unit square, cannot be expressed as rational numbers, which are fractions like $1/2$ or $1/3$. This discovery eventually led to the notion of *irrational* numbers that cannot be expressed as a ratio of integers."

Dorothy nods. "Yes, I know that the length of a diagonal of a square (with edges of length 1) is the square root of 2, or 1.4142135623. . . . This irrational length comes from the Pythagorean formula."

"Dorothy, you are a genius. Notice that the digits in the square root of 2 don't seem to follow any obvious pattern and that they continue forever. Similarly, the ratio of the circumference to the diameter of a circle is not a rational number. It is pi, or 3.1415926535. . . . Pi (written π) is also an irrational number. Its digits never end, and they don't form endless repeats like the digits of the rational number $1/7$ = 0.14285714285714285714. . . ."

"True."

"Dorothy, here is my unbelievably difficult question for you. If the digits of an of an irrational number are chosen at random, they surely would not be expected to display obvious patterns in the first one hundred digits. Is that right?

Mr. Plex jumps in. "No, that is not right. Consider the class of numbers that Earthlings call *zebra irrationals,*" Two of his feet started tapping excitedly. "And now I see why! Here is my favorite zebra irrational." Mr. Plex hands Dorothy and Dr. Oz a card with the equation:

$$f(n) = \sqrt{9/121} \times 100^n + (112 - 44n)/121$$

Mr. Plex looks into Dorothy's eyes. "I want to show you a beautiful irrational number produced when n equals 30." Mr. Plex hands her another card with the strange-looking number:

```
2727272727272727272727272727272727.
27272727272727272727272727272727
08
969696969696969696969696969696969
69696969696969696969696969696969
08
280134
680134  680134  680134  680134
680134  680134  680134  680134
676012928095772540216984661429105
73550317994762439206883650982326573720 74 . . .
```

Worlds' most wonderful zebra irrational

Dorothy backs up. "Oh my God, such patterns."

"Yes, I typeset it so you could see the patterns more obviously."

Dr. Oz nods, apparently impressed with Mr. Plex's mathematical prowess. "There are some bizarre patterns here, but suddenly they seem to stop at the last 680134, like water gushing from a hose and suddenly stopping when the supply is turned off. From there on, the digits follow no pattern that my alien eye can discern. Dorothy, here's a calculator. Can you compute more numbers of this zebra irrational and find any more patterns? What other zebras can you find?"

Difficulty Level: ✺ ✺ ✺ ✺

34 Creatures in Resin

"H. M.," said the Woggle-Bug, pompously, "means Highly Magnified; and T. E. means Thoroughly Educated. I am, in reality, a very big bug, and doubtless the most intelligent being in all this broad domain."

"How well you disguise it," said the Wizard. "But I don't doubt your word in the least."

– Dorothy and the Wizard in Oz

"**W**elcome to Oz II," Dr. Oz says as he points his tentacle at a large testing facility. To Dorothy, it appears to be a cluster of concrete domes, power plants, and manufacturing facilities of some kind – all partially buried in the Kansas prairies. Communication antennas sprout like weeds.

Robot squid are everywhere, digging, lifting, pushing; Oz II is growing like an ant colony.

"Glad you like the place," Dr. Oz says.

"I didn't say I like it."

Dr. Oz reaches for Toto.

"Okay, okay, I love Oz II."

Dr. Oz points to a large array of creatures trapped, or perhaps preserved, in some kind of polymer resin. "Here is your next test. You have five minutes to find the largest block of repeating creatures. I've highlighted two smaller repeating blocks in gray. Can you find any larger repeats?"

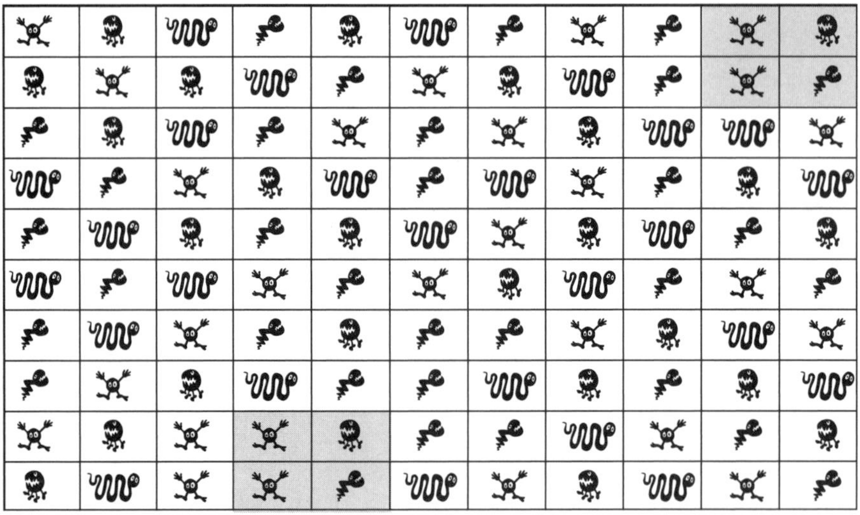

Creatures trapped in resin

Difficulty Level: 🎇

35 Prime-Poor Equations

The sole cause of man's unhappiness is that he does not know how to stay quietly in his room.

— Blaise Pascal, *Pensées* (1670)

Dorothy and Dr. Oz step through a concrete tunnel and into a vehicle shaped like a bullet. As they travel, Dorothy notices that the fluid-filled wheels continually change size to absorb the unevenness of the terrain. She feels as if she is riding across the land in a water-bed.

"Dorothy, recall that a prime is a positive integer, like 5 or 11, that is divisible only by itself and 1. Certain formulas are known to produce an astounding number of prime numbers." He punches a few buttons on the vehicles console, which then displays:

$$p = x^2 - x + c, \qquad c = 1, 2, 3, \ldots$$

Dorothy looks away from the formula and at their "car." Most surfaces of the vehicle are coated with a moist array of seaweed. It smells like miso soup. Perhaps this was Dr. Oz's food supply.

"Dorothy, pay attention. This is a famous prime-rich equation. Swiss mathematician Leonhard Euler (1707–83) discovered that for $c = 41$ there are forty values of p that are prime for an integer x, $1 \leq x \leq 40$. For example, for $x = 2$ and $c = 41$, we obtain $p = 43$, which is a prime number. If $x = 3$ and $c = 41$, we obtain 47, which is prime. Here are a few examples." Dr. Oz shows Dorothy a table of numbers:

x	p	x	p
1	41	7	83
2	43	8	97
3	47	9	113
4	53	10	131
5	61	11	151
6	71	12	173

Prime richness

Dorothy nods. "Impressive."

"However, much less is known about prime-poor equations of this form. Can you find a value of c for which the density of prime numbers is exceedingly low over some range? You have exactly one week to find such a value for c."

Difficulty Level: ✳ ✳ ✳ ✳

36 Number Satellite

B.C. MANSFIELD

The cowboys have a way of trussing up a steer or a pugnacious bronco which fixes the brute so that it can neither move nor think. This is the hog-tie, and it is what Euclid did to geometry.

— Eric Temple Bell, in R. Crayshaw-Williams, *The Search for Truth*

Dorothy and Dr. Oz are floating 100 miles from the surface of the Moon when they come upon a beautiful satellite with circles connected by roads or lanes. Each circle has three different roads.

"Dorothy, you are to choose a circle, and start your journey on any one of its three roads. Continue to travel through the roads. Each time your road goes through a circle you must add one or the other number that touches the road" (Figure 36.1).

B.C. MANSFIELD

36.1. Number satellite. (Illustration by Brian Mansfield.)

"Dr. Oz, could you give me an example?"

Dr. Oz points to a luminescent diagram on his flexscreen (Figure 36.2). "Yes, here's an example. If you start at the road toward the upper left, marked with the star (★), you could follow the path shown. You could start your running sum with either 100 or 101. The next road allows you to add either 7 or 6 to your sum. The next road allows you to add 7 or 999 to your sum. You are not allowed to travel a road in the circle more than once and you can't back up along a road in a circle."

36.2. Number satellite example. (Illustration by Brian Mansfield.)

"Okay." Dorothy studies the number satellite.

"Dorothy, your goal is to travel around the paths and return to the same circle you started from with a sum of 2003. If you do this within 36 hours, I will establish a permanent colony of trochophores on the Moon and name the Moon base after you."

Difficulty Level: ✷ ✷

37 Flatworm Math

In most sciences, one generation tears down what another has built, and what one has established another undoes. In mathematics alone each generation adds a new story to the old structure.

> — Herman Henkel, from S. Gudder's *A Mathematical Journey*

After spending a long September day surfing in cyberspace, Dr. Oz turns off his computer and decides to try for some real waves. The pond is not too cold, but this is probably his last swim of the season. There are enough small ripples that his surfboard actually moves a little in the wind.

In a few minutes, Dr. Oz is quickly surrounded by several dozen flatworms. They are at least a centimeter long, and as Dr. Oz tries to splash them away, he inadvertently severs a flatworm's body.

Coincidentally, Dr. Oz had that very day read about the flatworm's marvelous regenerative powers, so he knows that both pieces of the worm will probably live, each developing into whole. After gazing for a few minutes into the pond, Dr. Oz develops a problem called Flatworm Math, which, like the scenario in his pond, involves worm multiplication and severing. In particular, the problem consists of the repeated multiplication and truncation ("cutting") of integers, and Dr. Oz thinks many of you will enjoy simulating this, either on a computer or using a pencil and paper.

Start by picking any two-digit even number. Multiply by 2; sever the result, if there are more than two digits, by retaining only the last two; and multiply by 2 again. Repeat the process. Thus, for example, 12 becomes 24, then 48, then 96, then 92 ($2 \times 96 = 192$, and we chop off the leading 1), and so on. Below is a diagram indicating how the starting number 12 takes twenty steps to return.

$$12 \Rightarrow 24 \Rightarrow 48 \Rightarrow 96 \Rightarrow 92 \Rightarrow 84 \Rightarrow 68 \Rightarrow 36 \Rightarrow 72 \Rightarrow 44$$
$$\Uparrow \qquad\qquad\qquad\qquad\qquad\qquad\qquad\qquad\qquad\qquad \Downarrow$$
$$56 \Leftarrow 28 \Leftarrow 64 \Leftarrow 32 \Leftarrow 16 \Leftarrow 08 \Leftarrow 04 \Leftarrow 52 \Leftarrow 76 \Leftarrow 88$$

Flatworm sequence

Do flatworm sequences always return to their starting numbers? If so, how many steps are usually needed?

Difficulty Level: 🐛 🐛

38 Regolith Paradox

We have a habit in writing articles published in scientific journals to make the work as finished as possible, to cover up all the tracks, to not worry about the blind alleys or describe how you had the wrong idea first, and so on. So there isn't any place to publish, in a dignified manner, what you actually did in order to get to do the work.

— Richard Feynman, Nobel Lecture, 1966

Dr. Oz and Dorothy are traveling across a Kansas prairie and enter a small building built by Dr. Oz's fellow trochophores. The walls are made of shiny regolith that glistens like thousands of amethyst crystals in the morning sunlight. Ducts and cables dangle from the ceilings. Other trochophores ooze like wet worms over the walls.

"Dorothy, here is a new social hall we've created."

"Impressive," Dorothy says as she looks at the bars, restaurants, and trochophores eating segments of living sushi.

"Today I have a delightful problem for you involving some symbols from our language. Consider the three triplets on the wall."

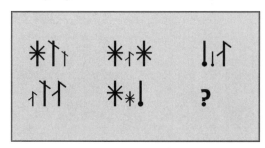

"Can you choose one of the following that best accompanies the above group?"

a b c

"We call this the Regolith Paradox," explained Dr. Oz.
"But what does this have to do with regolith?"
"Nothing. That's the paradox."

Difficulty Level: 🕸🕸

39

The union of the mathematician with the poet, fervor with measure, passion with correctness, this surely is the ideal.

— William James, *Collected Essays*

Dr. Oz is lecturing his friends (fellow trocophores, cloaked to appear human) in a small restaurant in a secluded area of Topeka, Kansas. He scrawled on a napkin the following numbers:

1, 2, 4, 5, 7, 9, 10, 12, 14, 16, . . .

After swallowing a spicy piece of meat, Dr. Oz turns to his friends. "If any of you can tell me the next number in this sequence, I will offer you a great reward."

After a few minutes, Dorothy calls out, "Sir, the next number is 17!"

Dr. Oz stands up. "Correct. I will buy you dessert today."

While waving his tentacles, Oz tells Dorothy about this weird and little-known sequence in number theory called the *Connell sequence.* "This sequence, proposed in 1959, is constructed by examining the consecutive natural numbers 1, 2, 3, 4, 5, . . . , and taking the first odd number, the next two even numbers, the next three odd numbers, the next four even numbers, and so on. Practical use of this sequence in antenna theory is discussed in a paper by the brilliant and enigmatic Indian mathematician Akhlesh Lakhtakia."

Dr. Oz begins to draw animal pictures on the napkin. "Here's an easy way to visualize what's happening. The bird 🐦 represents odd numbers, and Toto 🐕 represents even numbers. The sequence therefore looks like the following."

Dr. Oz considers Dorothy, who smiles back. "I have a second question," Dr. Oz says. "If you answer correctly, I will be your servant for a day." He pauses for dramatic effect. "How fast does this sequence grow? In other words, let u_n be the nth term in the sequence. Given the nth term, what is the ratio u_n/n?" Dr. Oz sketches a table of the first few entries:

n	u_n	u_n/n	n	u_n	u_n/n
1	1	1	5	7	1.4
2	2	1	6	9	1.5
3	4	1.3	7	10	1.43
4	5	1.25			

Dr. Oz looks at the rest of his friends. "Does the ratio grow increasingly large or is there some limit to its growth? Can you think of a formula that generates the nth Connell number? Is the one-millionth animal a Toto?"

Difficulty Level: 🐕 🐕 🐕 🐕

40 Entropy

Now one may ask, "What is mathematics doing in a physics lecture?" We have several possible excuses: first, of course, mathematics is an important tool, but that would only excuse us for giving the formula in two minutes. On the other hand, in theoretical physics we discover that all our laws can be written in mathematical form; and that this has a certain simplicity and beauty about it. So, ultimately, in order to understand nature it may be necessary to have a deeper understanding of mathematical relationships. But the real reason is that the subject is enjoyable, and although we humans cut nature up in different ways, and we have different courses in different departments, such compartmentalization is really artificial, and we should take our intellectual pleasures where we find them.

         — Richard Feynmann, *The Feynman Lectures on Physics*

Dorothy is with Dr. Oz a few miles away from the main Oz testing facility. Their hovercraft passes one of the largest buildings Dorothy has ever seen. It looks like an obelisk, perhaps a mile high and 30 feet in diameter. The adjacent Kansas prairie is scarred by large pits, evidently the result of some kind of mining operation. Around the amber obelisk are what look like furnaces enclosed by metallic cages.

"Dr. Oz, you are destroying Kansas's pristine environment."

"Don't worry about it. There are infinitely many parallel universes. If one Kansas gets a little dirty, there are infinitely many others that remain unaffected."

Dr. Oz leads Dorothy to a glass room. "Here is today's puzzle. There are 16 little clones of Mr. Plex bouncing around a room with 3 large Mr. Plex clones. Based solely on mathematical and physical assumptions about randomly moving particles, where do you expect the 3 large Mr. Plex's to be after a long time?"

Dorothy looks into the glass room. "Have you given me enough information?"

Dr. Oz rolls his eyes. "I always give you enough information."

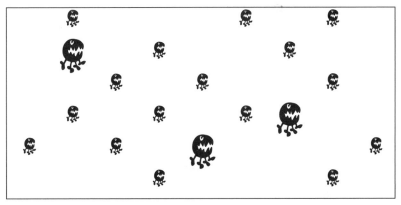

Bouncing clones of Mr. Plex. Where will they be after a long time?

Difficulty Level: ✳ ✳ ✳ ✳

41 Animal Gap

What I am going to tell you about is what we teach our physics students in the third or fourth year of graduate school. . . . It is my task to convince you not to turn away because you don't understand it. You see my physics students don't understand it. . . . That is because I don't understand it. Nobody does.

— Richard Feynman, *QED: The Strange Theory of Light and Matter*

D r. Oz's hovercraft passes a large prairie north of Wichita, Kansas. To the left is a large pyramid, a pile of rusted computer hard drives, and a small statue of a squid holding the hand of a humanoid. Small fabric flags surround the figure.

"What's that?" Dorothy says.

"It's a shrine to physicist Richard Feynman, one of our gods."

"Shrines, in the middle of a Kansas prairie?

"We come from far away. We've changed our attitudes, but our basic essence remains the same."

"What's that supposed to mean?"

"Not much."

Dr. Oz seems to be speaking in meaningless riddles. After traveling another mile, the hovercraft comes to an abrupt stop in front of a monolith covered with animals.

Dorothy steps out of the hovercraft. "I suppose this is my next intelligence test. Let me ask you a question. Why are animals missing near the middle?"

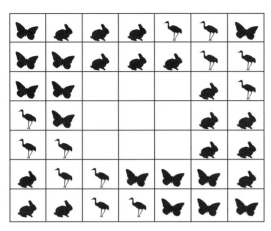

Animal monolith

Oz places his tentacle through the middle of the ring of animals and hands Dorothy a card. "Which of the three patterns fits into the empty section?"

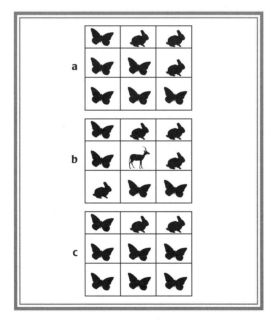

42 Arranging Alien Heads

Mathematical aesthetics seems more closely aligned with dance than with any time-invariant media.

— Clem Padin, *Science News*

Dr. Oz is with Dorothy in an autopsy room, staring at four alien heads on a stainless steel table. "Dorothy, how many ways can you group the four heads in a row?"

"Dr. Oz, we are not here to do math. We've got to figure out where these strange creatures came from. The information we find could affect the entire Earth."

"Dorothy, the answer is five. There are five different ways in which four heads can be grouped in a row. Here, let me show you." He starts to arrange the heads in different patterns.

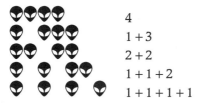

	4
	1 + 3
	2 + 2
	1 + 1 + 2
	1 + 1 + 1 + 1

Dorothy sighs. "That's all very interesting, but we've got work to do. Hand me a scalpel. I want to study their anatomy."

Dr. Oz continues. "The *partition function p(n)* is the number of distinct ways in which *n* can be written as a sum of smaller or equal integers. For example, 4 can be expressed in five different ways. Hence

$p(4) = 5$. The problem is equivalent to finding the number of different ways in which four objects can be grouped in a row."

"Get out of here," Dorothy yells at Dr. Oz as she throws a head at him.

* * *

The rest of this chapter describes a simple graphics technique allowing you to visualize partitions of integers. *Warning:* If you are not interested in technical mathematics, you may wish to skip this chapter. For those brave souls who remain, Dr. Oz will try to help you understand the "theory of partitions." Pick an integer, for example, 4. As you just learned, "partitions" of 4 specify the number of ways you can sum integers to produce 4. For example, $p(4) = 5$ since $(4 = 4)$, $(4 = 1 + 3)$, $(4 = 1 + 1 + 2)$, $(4 = 1 + 1 + 1 + 1)$, and $(4 = 2 + 2)$.

Given the integer a, we say that a sequence of positive integers $n_1, n_2, \ldots, n_r$ $(n_1 \leq n_2 \leq \ldots \leq n_r)$ is a *partition* of a if $a = n_1 + n_2 + \ldots + n_r$. Let $p(a)$ denote the number of partitions of n. In the previous paragraph Dr. Oz suggested why $p(4) = 5$. Dr. Oz is interested in the more specific problem of finding *all* series of *consecutive* positive integers whose sum is a, where a is a positive integer. (Dr. Oz seeks a partition of a into consecutive integers.)

As an example, there are four partitions of 10,000 into consecutive integers: $[18 + 19 + \ldots + 142]$, $[297 + 298 + \ldots + 328]$, $[388 + 389 + \ldots + 412]$, and $[1998 + 1999 + \ldots + 2002]$. $P_c(a) = 4$. Do most numbers less than 10,000 have four partitions into consecutive integers?

Figure 42.1 is a graphical representation of the distribution of all such sequences for $1 \leq a \leq 200$. The x-axis contains the a values, and the runs of dots along the vertical y direction indicate the partitions for these a values. (*Note:* This is not a plot of a versus $P_c(a)$ but a plot of a versus all consecutive partitions of a.) The graph shows a number of interesting features. For example, notice, as expected, the dots never exceed $\frac{1}{2}(a + 1)$ because the largest value possible in a series of consecutive positive integers whose sum is a must be $\frac{1}{2}(a + 1)$. Perhaps the most apparent feature on the graph is the rising (diagonal) two-dot cluster at the top of the graph. These dots correspond to odd values of a that always have $[\frac{1}{2}(a + 1)] + [\frac{1}{2}(a + 1) - 1]$ as a partition. For example, $21 = \sum_{10}^{11} n$. Prime-number values for a have this kind

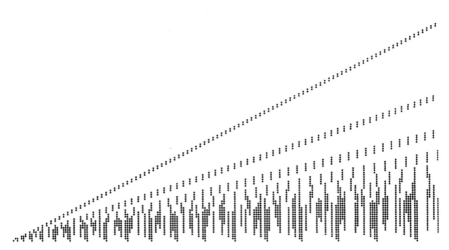

42.1. Partition of *a* into consecutive integers. This graphical representation shows the distribution of the partitions for $1 \leq a \leq 200$ on the *x*-axis. The *y*-axis goes from 0 to 100.

of binary partition and no others. The diagonal line of triplet-dots in the graph represents values of *a* evenly divisible by 3, which always have as a partition $(n/3 - 1) + n/3 + (n/3 + 1)$. Interestingly, magnification of the graph reveals gaps where no partitions into consecutive integers exist. These occur at $a = 2^m$, $m = 1, 2, \ldots$. Dr. Oz urges you to find other interesting properties of the partitioning of *a* into consecutive integers by studying the graph.

What other related observations can we make about partitions?

Difficulty Level: ✺ ✺ ✺ ✺

43 Ramanujan Congruences and the Quest for Transcendence

An equation for me has no meaning unless it expresses a thought of God.
— Srinivasa Ramanujan (1887–1920)

Dr. Oz leads Dorothy into a cluster of huts. In the back of one is a small fission power plant with tanks of water and bubbling gasses. Robot squid crawl between the equipment, monitoring the blinking lights and dials, their tentacles stained by chemicals Dorothy had no desire to smell.

"Dorothy, in the previous chapter we talked about partitions of integers."

"Dr. Oz, what do you mean by 'chapter?' You act as if we're in some kind of book."

"Oh, I meant to say that yesterday we talked about partitions of integers. Partitions count the number of ways a particular integer can be expressed as the sum of positive integers. For example, 5 can be written as follows." Dr. Oz hands Dorothy a card with the different ways of writing 5:

$$5$$
$$4 + 1$$
$$3 + 2$$
$$3 + 1 + 1$$
$$2 + 2 + 1$$
$$2 + 1 + 1 + 1$$
$$1 + 1 + 1 + 1 + 1$$

"We don't count different arrangements of the same integers. For example, $3 + 2$ is the same as $2 + 3$. You can see there are seven ways of representing 5. A year before his death in 1920, mathematician Srinivasa Ramanujan was studying a list of partitions for the numbers 1 to 200. (The number of partitions ranged from 1 to 3,972,999,029,388.)"

Dorothy puts her fingers over her nose. For some reason, the odors are quite strong in Dr. Oz's facility. "Sir, there are seven partitions for 5. How many are there for slightly bigger numbers?"

"I'm glad you asked. The partitions of n for n ranging from 1 to 34 are: 1, 2, 3, 5, 7, 11, 15, 22, 30, 42, 56, 77, 101, 135, 176, 231, 297, 385, 490, 627, 792, 1002, 1255, 1575, 1958, 2436, 3010, 3718, 4565, 5604, 6842, 8349, 10,143, 12,310. Ramanujan noticed that, starting with 4, the number of partitions for every fifth number is a multiple of 5." Dr. Oz hands Dorothy a card showing the numbers that exhibit this behavior:

1, 2, 3, *5*, 7, 11, 15, 22, *30*, 42, 56, 77,
101, *135*, 176, 231, 297, 385, *490*, 627,
792, 1002, 1255, *1575*, 1958, 2436,
3010, 3718, *4565*, 5604, 6842, 8349,
10,143, *12,310*

"Dorothy, this observation can be expressed in a mathematically compact form, where $p(n)$ is the partition function of n. This particular congruence would be stated as shown on the card in my tentacle."

$$p(5N + 4) \equiv 0 \text{ (mod 5)} \text{ for } N \geq 0$$

"Here the triple-line symbol ≡ indicates 'congruence,' and 0 (mod 5) means that the remainder after division by 5 must be 0." He pauses to flip her another card. "Notice that starting with 5, the number for every seventh integer is a multiple of 7."

1, 2, 3, 5, *7*, 11, 15, 22, 30, 42, 56, *77*,
101, 135, 176, 231, 297, 385, *490*, 627,
792, 1002, 1255, 1575, 1958, *2436*,
3010, 3718, 4565, 5604, 6842, 8349,
10,143, 12,310

"We can write this in the following mathematical notation."

$$p(7N + 5) \equiv 0 \text{ (mod 7)} \text{ for } N \geq 0$$

"Starting with 6, the number of partitions for every 11th integer is a multiple of 11."

1, 2, 3, 5, 7, *11*, 15, 22, 30, 42, 56, 77,
101, 135, 176, 231, *297*, 385, 490, 627,
792, 1002, 1255, 1575, 1958, 2436,
3010, *3718*, 4565, 5604, 6842, 8349,
10,143, 12,310

"We can write this in the following mathematical notation."

$$p(11N + 6) \equiv 0 \ (\text{mod } 11) \ \text{for } N \geq 0$$

"Similar relations have been found when integers are powers of 5, 7, or 11, or a product of these powers. Ramanujan was able to prove that the patterns highlighted here hold for all higher numbers, not for just the first few."

"Dr. Oz, this is pretty interesting. Where are you heading with this line of thought?"

"What was amazing about Ramanujan's discovery is that there is nothing in the definition of partitions that would suggest such relationships, which are known to mathematicians as 'congruences.' Also, we can still wonder why prime numbers like 5, 7, and 11 should be the ones that produces the congruences."

Dr. Oz looks Dorothy directly in the eye. "Dorothy, here is your challenge. Can you find any other congruences?"

Difficulty Level: ✺ ✺ ✺ ✺

44　Getting Noticed

You're drifting into deep water, and there are things swimming around in the undertow you can't even conceive of.

— Stephen King, *Insomnia*

Dr. Oz points a ray gun at Dorothy and fires. She is suddenly transformed into an ant. Dr. Oz leaves her alone to survive.

What is the best way for Dorothy to communicate to humans that she is really still an intelligent human trapped in an ant's body, without being squashed first? For example, she might spell out HELP with rice grains, or get other ants to follow her to spell words like:

PLEASE HELP ME

Although this approach seems difficult to implement, Dorothy might be able to at least arrange dead ant bodies in various configurations. Perhaps she should spell out some mathematical or physics formula, but which one? Should she spell out Einstein's famous one?

E = MC2

What message would you write? What else could you do?

Difficulty Level: 🐜

45 Juggler Numbers

Some people can read a musical score and in their minds hear the music. . . . Others can see, in their mind's eye, great beauty and structure in certain mathematical functions. . . . Lesser folk, like me, need to hear music played and see numbers rendered to appreciate their structures.

— Peter B. Schroeder, *Byte*

One spring morning, while watching a juggler throw yin-yang-patterned balls through the air of a circus tent, Dr. Oz developed a seemingly simple number sequence. Like a ball thrown up and down by a circus juggler, this sequence also drifts down and up, sometimes in seemingly haphazard patterns. Also like a juggler's ball, the Juggler sequence apparently always falls back down to the juggler's hand (which we represent by the integer "1").

Start with any positive integer. If it is even, raise it to the $\frac{1}{2}$ power (i.e., simply take the square root of the number). If it is odd, raise it to the $\frac{3}{2}$ power (i.e., take the square root of its cube). The computational recipe can be represented as follows:

```
Input positive integer x
repeat
    if x is even
        then x ← [x^(1/2)]
        else x ← [x^(3/2)]
until x = 1
```

In either case, take your result and truncate it (the square brackets) to the maximum integer equal to or smaller than the result (e.g., 4.1 becomes 4). Repeat the operations over and over again. This sequence is produced by iterative rules: Apply the rule to the current number in the sequence and you get the next one.

Dr. Oz turns to Dorothy. "What is the long-term behavior of such sequences? Will they settle into a cycle? If so, what sort of cycle? If you can program your computer to calculate fractional powers (and you can), you will be able to create a Juggler sequence. In your program you can continue the iteration process until the Juggler number returns to 1."

Dorothy ponders the barrage of questions. Was Dr. Oz foolish to terminate the loop with "until $x = 1$"? What if the sequence never returns to 1?

Apart from this nasty little question, the more general question of what patterns evolve is sure to delight. Consider for example what happens when the process goes to work on the number 3 as input: 3, 5, 11, 36, 6, 2, 1. We can represent the rise and fall of this sequence as follows:

The basic pattern is simple enough. It rises for several steps and then gradually falls to 1. In this particular example, the sequences of 🌑's is "bipartite" because the numbers (symmetrically) rise to a peak and fall, in this case, three steps up and three steps down. (How common are bipartite Juggler sequences?) Once the value goes to 1, the value repeats: 1, 1, 1, ...

The Juggler problem is an interesting variation of the famous $3n + 1$ (or "Hailstone" or "Collatz") problem discussed extensively in the mathematical literature. Hailstone sequences evolve according to the rule

```
if x is even
    then x ← x/2
    else x ← 3x + 1
```

Juggler sequences use an analogous rule involving *powers* instead of multiplications and divisions. Like the Hailstone sequence, the Juggler sequence drifts down and up, sometimes in seemingly chaotic patterns. You are forewarned that, unlike the Hailstone sequence, Juggler numbers can reach amazingly huge values in just a few iterations. However, even though large numbers can be reached rather rapidly (for example, 24,906,114,455,136 is the 9th member of the Juggler sequence that starts at 37), it seems that even these large values soon decay fairly quickly to 1. One may conjecture that, starting from any positive integer, repeated iteration of this function eventually produces the value 1. Dr. Oz has checked this conjecture for the Juggler sequence for all starting numbers below 200. They all do converge to 1, but fairly long sequences can be generated by the Juggler procedure before the values settle back to 1. Would any of you like to test Dr. Oz's conjecture beyond 200? To get you in the mood, here is an example for the starting number 77:

77, 675, 17,537, 2,322,378, 1523, 59,436, 243, 3787, 233,046, 482, 21, 96, 9, 27, 140, 11, 36, 6, 2, 1

This has a path length of 20 steps. You can see that the pattern here is not so simple as the previous example of 3. Here there are six ups and six downs in a haphazard order.

Figure 45.1 shows the path length of the sequence for starting integers between 0 and 175. The up-and-down zigzags in the line indicate highs and lows corresponding to consecutive odd and even starting integers. Usually the sequence is fairly tame, reaching its maximum value in a few steps, and returning to 1 in under 10 iterations. There are some notable exceptions: The starting numbers 37, 77, 103, 105, 109, 111, 113, 115, 129, 135, 163, 165, 173, 175, 183, 193, and others give fairly long sequences. Why should that be? Also note that if a number of the form 2^{2^n} (e.g., 4 or 16), then the sequence must monotonically decrease to the ground state, 1.

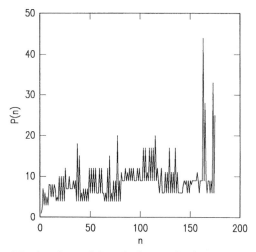

45.1. Juggler path lengths for starting integers between 0 and 175.

Dr. Oz has also observed that, as far as the first 200 integers are concerned, only three terminating three-digit patterns occur: (6 2 1), (4 2 1), or (8 2 1). Why only these three? Figure 45.2 shows monotonic end sequences, called *decay paths,* observed for starting values less than 200.

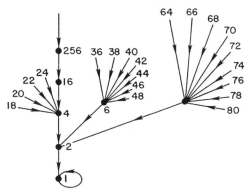

45.2. Monotonic decay paths. These are for the end Juggler sequences observed for starting values less than 200. The central trunk of the tree comprises numbers of the form 2^{2^n}.

Dr. Oz invites you to extend the search for longer Juggler sequences. In addition, you can experiment with modified Juggler sequences determined by rounding to the nearest integer (rather than simply rounding down). Whether or not the Juggler sequences or modified Juggler sequences always settle back to 1 is not known. This is a fiendishly difficult challenge to resolve. You may find that the Juggler problem is of interest because it is so simple to state yet intractably hard to solve.

Do all Juggler numbers return to 1?

Difficulty Level: 🐜🐜🐜🐜

46 Friends from Mars

Your vision will become clear only when you can look into your own heart. Who looks outside dreams; who looks inside, awakens.

— Carl Jung

Dr. Oz is showing some of his alien friends around Las Vegas, Nevada.

 is Mr. Plex.

 is a Venusian named Violet.

 is a Jovian named Jake.

Each of the creature's enclosures must be connected to its twin enclosure with an umbilical cord in order for the creature to survive. Your job is to connect the cells containing the creatures with cords that do not cross or go outside the surrounding walls. Your cords must be along the floor. They may be curvy, but they cannot touch or cross one another. You can't draw lines through the cells surrounding each alien.

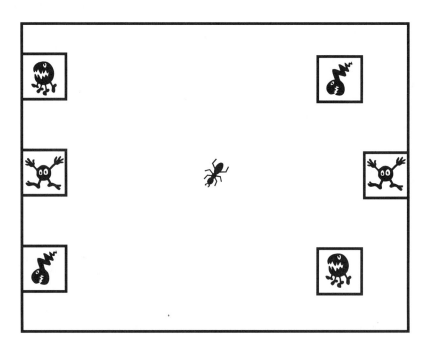

Most people who can't solve the puzzle can solve it if they put it away for a day and then look at it again.

Difficulty Level: 🦟

47 Phi in Four 4's

Mathematics is the wrong discipline for people doomed to nongreatness.
— Don DeLillo, *Ratner's Star*

r. Oz and Dorothy are in a cave, facing a subterranean lake. The lake appears clear in the light of a flashlight. Dorothy blows on the water, and countless ripples appear on its surface. Beneath the water she sees quartz nodules the size of grapefruits.

Dorothy closes her eyes for a second. Even the faint dripping sound is music. She has longed for this fragrant and embraceable coolness. Was there anyplace else in Kansas, the world, where the moist air felt like a living presence, where the cold air brushes against you like cats, where the stalactites and stalagmites and slippery cave walls are shimmering and alive?

Can Dr. Oz feel what she is feeling? Could he appreciate the majesty and mystery of this Stygian chamber? He seems deep in thought.

"What are you thinking about?" Dorothy says.

"About mathematics, naturally. The number 1.61803..., called the golden ratio, appears in the most surprising places, and because it has unique properties, mathematicians have given it a special symbol." On the cavern floor he draws a big

$$\phi$$

"This symbol is the Greek letter phi, the first letter in the name Phidias, the classical Greek sculptor who may have used the golden ratio in his work. A golden *rectangle* has a ratio of the length of its sides equal to $1 : \phi$. Some of you humans have reported that the golden rectangle is among the most visually pleasing of all rectangles, being neither too squat nor too thin." He sketches a golden rectangle on the ground:

Golden rectangle

Dr. Oz goes on to explain that because $\phi = (1 + \sqrt{5})/2$, it has some rather amazing mathematical properties. For example,

$$\phi - 1 = 1/\phi \qquad \phi\Phi = -1 \qquad \phi + \Phi = 1 \qquad \phi^n + \phi^{n+1} = \phi^{n+2}$$

where $\Phi = (1 - \sqrt{5})/2$. Both ϕ and Φ are the roots of $x^2 - x - 1 = 0$. It is this famous constant ϕ that is known as the *golden ratio*.

"Dorothy, here is your puzzle for today. Arrange four 4's, and any of the ordinary mathematical symbols, to give as close an approximation to $\phi = 1.61803$ as you can find. Allow yourself the following symbols:), (, +, −, ×, ÷, as well as the usual notations for roots, powers, factorials, and the decimal point. Factorials are to be of integers only. Concatenation of 4's is allowed (e.g., 44). For *Contest 1,* use as many mathematical symbols as you wish. For *Contest 2,* limit yourself to using at most four of each mathematical symbol. For example, you can only use the multiply symbol four times.

Dorothy ponders the question for a second. "Give me just one example."

"Okay, here is one approximation with four 4's. Can you find others? Can you find more accurate ones?" On the ground he draws:

$$\sqrt{\sqrt{\sqrt{44/(4/4)}}} = 1.6048394$$

Difficulty Level: ✳ ✳ ✳

48 On Planet Zyph

It is a mathematical fact that the casting of this pebble from my hand alters the center of gravity of the universe.

— Thomas Carlyle, *Sartor Resartus III*

"**D**orothy, on our home world, which we call planet Zyph, grows a Zyph-berry tree. Each branch contains five circular berries divided into four quadrants." Dr. Oz points to a drawing of a black and white tree with circular fruits. They remind Dorothy of little stained-glass windows, as bright light shines through the white portions of the berries (Figure 48.1).

"Here is a Zyph tree with five branches. One branch does not belong. It is a fraud, a hoax, attached to the tree by a devious fellow. Can you tell which branch does not belong and why? The only way you can tell is by superimposing the berries."

48.1. Zyph-berry tree. (Illustration by Brian Mansfield.)

Difficulty Level: ✹

49 The Jellyfish of Europa

"Can you do addition?" the White Queen asked. "What's one and one and one and one and one and one and one and one and one and one?"
"I don't know," said Alice. "I lost count."
— Lewis Carroll, *Through the Looking Glass*

Dorothy and Dr. Oz stand before an immense aquarium perhaps 10,000 gallons in volume. Strange jellyfishlike creatures frolic and prance, their diaphanous tentacles trailing behind them in a glittering, ancient sea. The jellyfish seem to have created a mazelike structure from the coral encrusting the aquarium walls (Figure 49.1).

"Dorothy, swimming in the seas of Europa, one of Jupiter's moons, are beautiful jellyfish that live in intricate mazelike cities. We have taken a number of specimens and placed them in this tank. Your mission is to swim through their city, collect all the jellyfish, and leave."

Dorothy watches as the jellyfish align themselves within the maze.

"The jellyfish are crafty. They have stinging cells. You must be careful. This means you must find a path that enters the maze, passes through all the jellyfish, and exits the maze without using any part of a path more than once. Not only can't your path cross itself, it also can't go through the same intersection more than once. Because the jellyfish will see you if you approach them from the front, you must collect them while passing from their backs to their fronts."

Dorothy nods as she dons scuba gear.

"Dorothy, how many different solutions can you find?"

49.1. Jellyfish of Europa. (Illustration by Brian Mansfield.)

B.C. MANSFIELD

Difficulty Level: ✷ ✷

50 Archaeological Dissection

Alice laughed: "There's no use trying," she said; "one can't believe impossible things."

"I daresay you haven't had much practice," said the Queen. "When I was younger, I always did it for half an hour a day. Why, sometimes I've believed as many as six impossible things before breakfast."

— Lewis Carroll, *Alice in Wonderland*

Dorothy and Dr. Oz have transported themselves to the Earth's largest pyramid, and the largest monument ever constructed – the Pyramid of Quetzalcoatl at Cholula de Rivadabia, 63 miles south of Mexico City. It is 177 feet tall and its base covers 45 acres.

Dorothy stops suddenly when she see a cross-shaped monolith covered by some dirt at the base of the pyramid (Figure 50.1).

"Dorothy, this is a puzzle for you. I want you to cut the symmetrical cross into five pieces so that, when viewed from the top, one piece will be a smaller symmetrical cross, and the remaining four pieces will fit together and form a perfect square. If you can perform such a dissection within an hour, I will grant all your wishes."

Dr. Oz hands Dorothy a chisel, and she wonders how to cut the monolith according to the directions (Figure 50.2). Where should she cut?

Difficulty Level: 🎇 🎇 🎇

50.1. Archaeological dissection. (Illustration by Brian Mansfield.)

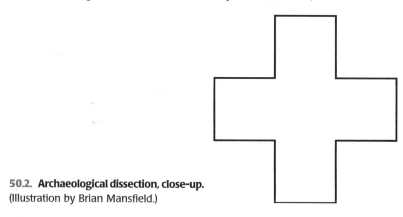

50.2. Archaeological dissection, close-up.
(Illustration by Brian Mansfield.)

51 The Gamma Gambit

"You should say what you mean," the March Hare went on.

"I do," Alice hastily replied; "at least I mean what I say, that's the same thing, you know."

"Not the same thing a bit!" said the Hatter. "Why, you might just as well say that 'I see what I eat' is the same thing as 'I eat what I see!'"

— Lewis Carroll, *Alice in Wonderland*

Dorothy and Dr. Oz are in a glass sphere, twenty feet in diameter, suspended five feet above the pavement of Highway 140, which straddles the counties of Saline and Ellsworth. She challenges Dr. Oz in guessing the date of a coin, any U.S. coin, that he is to bring at random from the iridescent pouch holding his newfound coin collection. (He'd recently captured a piggybank for experimentation, having mistaken porcelain for porcine.)

"Dr. Oz, I call this little puzzle the 'gamma gambit.'"

"And why do you call it that?"

"Just because gamma is a Greek letter, and I thought that would make the puzzle be more impressive to you than if I simply called it a guessing game.'"

"Proceed."

"We are to guess the date of the coin. I will wager that my guess will be closer to the date than your guess. You will have only one guess, while I have two, but to make up for my slight advantage, not only will I let you guess first, but I will wager you my wealth and my life if I lose. However, if I win, you must let me return to Aunt Em and Uncle Henry for a week's vacation from your testing me."

"Okay, sounds fair. Only after the guesses have been made will I bring the coin into view."

What strategy should Dorothy use to dramatically increase her chances of winning?

Difficulty Level: 🪲

52 Robot Hand Hive

The mind is like an iceberg; it floats with one-seventh of its bulk above water.

— Sigmund Freud

Dorothy looks out over the prairies at a trocophorean launch station. The buildings are burned out and abandoned. The roads are coated with a layer of gray sand.

"What happened here?"

Dr. Oz looks out over the desolate landscape. "The robotic hands got loose. We finally had to exterminate most of them. Take a look at this. It's your next puzzle." He points to a colony of disembodied robotic hands.

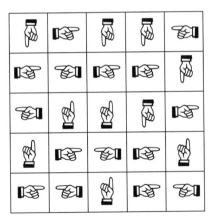

Robotic hand hive

"The hands have an unusual security system for their hive. They position themselves in square cells so that invaders who approach directly into their extended index fingers will run into their sharp, deadly nails. Can you find a path that will take you through the robot hive so that you go through every cell without encountering an index finger head-on? You must enter the hive at any perimeter cell, then move one cell at a time horizontally or vertically (never diagonally) and exit at any square. Go slowly! One false turn and you are impaled by a razor-sharp nail."

Dorothy puts her hands on her hips. "I'm sick of your exaggerations and continual reneging on my rewards!"

"Ah, but you'll never know when I'm being dead serious."

Difficulty Level: ✺ ✺ ✺

53 Ramanujan and the Quattuordecillion

In a perfect universe, infinities turn back on themselves.

— George Zebrowski, *OMNI*

"**W**ho is that?" Dorothy says pointing to a tall trocophore who is climbing out of a personal flying saucer. An array of robotic insects hover about the squidlike being's face. Perhaps the robots serve a defensive purpose, but Dorothy is not sure.

"She is one our high priestesses," Dr. Oz says. "Her name is Eve."

Eve waves to Dorothy. The priestess is wearing a black cassock and a necklace of Pentium IV computer chips. "Dorothy, I have always been fascinated by the mathematical concept of nested roots."

Dorothy bows before the priestess. "Oh Great One, I am not too sure about what Thou means."

"You are probably familiar with the commonplace square-root symbol in mathematics. Here is an example." Eve sketches in the soil with her posterior tentacle.

$$\sqrt{4} = 2$$

"Nested roots are certainly more fascinating (and typographically attractive) to look at. Here is an example."

$$\sqrt{\sqrt{4}} = 1.414\ldots$$

"If mathematics were a beauty contest, then *nested roots* would win the prize. They're simply square roots of roots of roots in an infinite cascade. Here's an example of an infinitely nested root."

$$\sqrt{2 + \sqrt{2 + \sqrt{2 + \ldots}}}$$

Dorothy nods. "That's a very attractive looking expression."

"Yes it is. A few years ago I asked colleagues if it were possible for humanity to determine the solution to the following nested set of square roots." Eve draws the following equation and places a box around it.

$$\boxed{\sqrt{1 + \heartsuit \sqrt{1 + (\heartsuit + 1) \sqrt{1 + (\heartsuit + 2) \ldots}}} = ?}$$

What is the value of $\heartsuit$?

"Ooh!" Dorothy steps back in awe. "Mon Dieu. What is that cool-looking heart symbol?"

"Here, $\heartsuit$ indicates a quattuordecillion, or 1 followed by 45 zeros."

$$\heartsuit = 1,000,000,000,000,000,000,000,000,000,000,000,000,000,000,000$$

Dr. Oz nods. "Yes, 'quattuordecillion' is the official name for this number in America. It comes from the Latin root *quattuordecim*. Check it out in your *Webster's Unabridged Dictionary*. To get a feel for the size of a quattuordecillion, it's much larger than the number of molecules in a pint of water (1.5×10^{25}) but fewer than the number of electrons, protons, and neutrons in the Universe (10^{79})."

Eve raises her tentacle and immediately silences Dr. Oz. "When I first became enamored by the number quattuordecillion I searched your world for large-number comparisons so that I could adequately convey to students the majesty of the quattuordecillion. Here are some other comparisons. The *Ice Age number* (10^{30}) is the number of snow crystals necessary to form the Ice Age. The *Coney Island number* (10^{20}) is the number of grains of sands on the Coney Island beach. The *talking number* (10^{16}) is the number of words spoken by humans since the dawn of time. It includes all baby talk, love songs, and congressional debates. This number is roughly the same as the number of words printed since the Gutenberg Bible appeared. The amount of money in circulation in Germany at the peak of inflation was 496,585,346,000,000,000,000 Marks. This number is *approximately* equal to the number of grains of sand on the beach at Coney Island. The number of oxygen atoms contained in the average thimble is a good deal larger: 1,000,000,000,000,000,000,000,000.

"Eve," Dorothy says, "you are a fountain of fascinating facts."

"Thank you. Here are some more. The number of electrons that pass through a filament of an ordinary light bulb in a minute equals the number of drops of water that flow over Niagara Falls in a century. The number of electrons in a single leaf is much larger than the number of pores of all the leaves of all the trees in the world. The number of atoms in a book is less than a *googol* (10^{100}, or 1 followed by 100 zeros). The chance that a book will jump from the table into your hand is not 0 – in fact, using the laws of statistical mechanics, it will almost certainly happen sometime in less than a googolplex of years ($10^{10^{100}}$, or 10^{googol}, or 1 followed by a googol zeros).

Dorothy smiles. "Eve, you're amazing."

Dr. Oz lets out a great gush of air. Perhaps he is jealous of Dorothy's adoration of Eve.

"Let's return to the heart equation in the box," Eve continues. "When I asked humans to determine a value for the infinite set of nested roots, I surely felt that none of you could arrive at a solution without a computer, and, even with a computer, I thought most people would have extreme difficulty. The behavior of the nested root also makes it particularly difficult to decipher. The continued roots seem to make the values smaller, while the $\heartsuit$ multipliers continue to make the values larger. Think of two sumo wrestlers competing for domination on a numerical wrestling mat. How could we ever arrive at an amicable solution?"

"Eve, I don't know what the answer is for the equation."

"This is your challenge. You have one month to give me an answer."

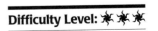

Difficulty Level:

54 The Lunatic Ferris Wheel

Oh, set me free. Let's speed on wheels, on wheels, on wheels. . . .

— Billy Idol, "Speed"

Dr. Oz is with Dorothy in his subterranean laboratory. "I have an idea," he says, "for the most wondrous Ferris wheel. It consists of three wheels of different sizes. The rider sits on a little wheel mounted on a bigger wheel that is mounted on a bigger wheel, each turning at a different rate." Dr. Oz draws a diagram on the board (Figure 54.1).

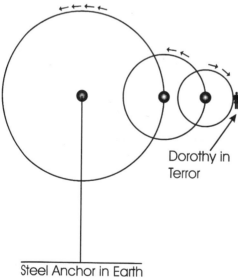

54.1. **The Lunatic Ferris Wheel.**

Dorothy looks at the Ferris wheel. "I think I'd get dizzy on this one."

Dr. Oz nods. "Perhaps. Now, let's leave the lab. I have a surprise for you."

They travel through a long tunnel to Earth's surface and emerge in a field of waving wheat. Dr. Oz smiles at his creation. "I actually built the darn thing. You will be my first passenger."

Dorothy stares at the large metal contraption. "Are you sure this is safe?"

"Certainly. I want you to sit on the chair and think about what paths you might trace as your body moves through the air. In this experiment, the largest wheel turns slowly counterclockwise. The second wheel travels seven times as fast. The small wheel turns 17 times as fast as the first wheel, clockwise, and out of phase."

"What about the wheel sizes?"

"The radius of the second wheel is half the first. The radius of the third wheel is one-third the first."

After an hour, Dorothy gets off the Ferris wheel. She brushes her hair out of her face. "I think I'm going to be sick." Her voice is hoarse from all her screaming.

"Yes, and I know why." Dr. Oz holds up a piece of paper with a drawing (Figure 54.2). "This is the wild path you traced out as you traveled on my Ferris wheel. Isn't it amazing that it has sixfold symmetry like a snowflake?"

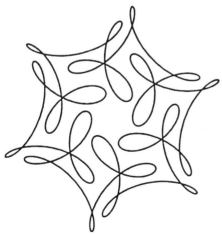

54.2. Dorothy's dizzying path along the Lunatic Ferris Wheel, the ride from hell. (After Frank Farris, "Wheels on Wheels on Wheels.")

"Amazing."

"Dorothy, I'm interested more generally in the paths you can trace out on the small wheel. What equations can you derive to calculate the motion of the rider through space? What other kinds of symmetrical drawings can the Lunatic Ferris Wheel produce?

Difficulty Level: 🐜🐜🐜🐜

55　The Ultimate Spindle

"Good day, mother," said the Princess, "what are you doing?"
"I am spinning," answered the old woman, nodding her head.
"What is that that twists round so briskly?" asked the maiden, and taking the spindle into her hand she began to spin.
　　— Jacob Ludwig Grimm and Wilhelm Carl Grimm, *The Sleeping Beauty*

The trochophore base outside of Wichita seems as if it is stranded in a endless prairie, connected to the Oklahoma border by a single, aging rail line. To the left are hangars and launchpads of various dimensions. Fuel tanks jut from the dry earth like ancient obelisks.

Dr. Oz approaches Dorothy holding some strange graphs in his hand. "Today I'm interested in the graph *x* to the *x*. It has the power to create interesting spindle shapes." He hands her a card:

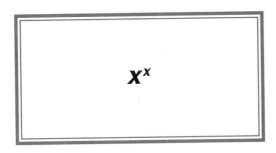

$$x^x$$

He then looks at his pocket viewscreen, which displays Figure 55.1.

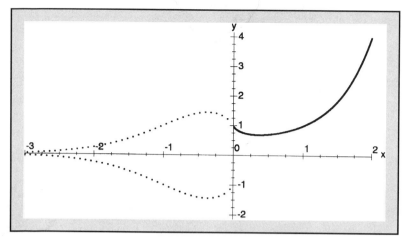

55.1. The bizarre graph of $y = x^x$ for real values of x and y. For $x < 0$, negative integer multiples of $\frac{1}{25}$ are plotted. (Figure courtesy of Mark D. Meyerson.)

"You can see that the graph is smooth for values of x greater than zero."

"Smooth as a baby's bottom," Dorothy says. She punches numbers into her calculator to generate a few values:

x	x^x
1	1
2	4
3	27
4	256
5	3,125
6	46,656
7	823,543
8	16,777,216

Rapidly growing x^x

"It grows rather fast," she says, "and you've just shown the curve for small values of x. But hold on. Why is the curve so ghostlike for x less than zero? Why is it broken up? And why do you say it can create spindles? Spindles are 3-D objects."

Dr. Oz taps his tentacle on a rusting fuel tank. "If you can answer those questions, I will set you free."

Difficulty Level: ✯ ✯ ✯ ✯

56 Prairie Artifact

Through space the universe grasps me and swallows me up like a speck; through thought I grasp it.

— Blaise Pascal, *Pensées* (1670)

While exploring a secluded Kansas prairie, Dorothy and Dr. Oz come upon an alien artifact with some odd-looking symbols:

Alien artifact

Dr. Oz caresses the artifact, and it speaks in a voice resembling newsman Dan Rather's. "You can see that there are five vertical pairs of symbols. You are to find a pair of symbols to complete the set from among the five possible solutions shown here."

Choose one solution

"This is a difficult problem!" Dorothy yells at the artifact.

"Yes, it is. If you do solve this puzzle, legend has it that you will be granted the power of invisibility that you can turn on and off at will."

Difficulty Level: ✹ ✹ ✹

57 Alien Pellets

Nature shows that with the growth of intelligence comes increased capacity for pain, and it is only with the highest degree of intelligence that suffering reaches its supreme point.

— Arthur Schopenhauer, *Parerga and Paralipomena*

A thirty-foot-tall winged alien flies to Dorothy and excretes *969* pellets the color of emeralds. Dorothy is not certain if she should be disgusted or delighted by the sparkling exudate.

**Winged alien excreting
969 emerald-color pellets**

Dorothy decides it would be prudent to back away, and the creature drops 486 pellets, then 192 pellets, and finally 18 pellets. The process is accompanied by horrendous sounds, like a cross between cannon blasts and moose calls. But perhaps the creature is trying to learn how smart Dorothy is. The alien stares at her with his huge dark eyes. He is waiting for Dorothy to determine how many pellets should come next. Hurry! If she doesn't answer within five minutes, the alien will produce gaseous blasts stinking of sulfur dioxide.

Difficulty Level: 🐦 🐦

58 The Beauty of Polygon Slicing

If we look into ourselves we discover propensities which declare that our intellects have arisen from a lower form; could our minds be made visible we should find them tailed.

— W. Winwood Reade

The coelenterates from Venus lived at the edge of a Kansas launch complex in temporary camps surrounded by barbed-wire fences. These creatures spent their time praying, protesting, and leafleting – all the while monitored by Dr. Oz's kin and his drone robots.

"Never mind them." Dr. Oz says. "Earth is hot real estate these days. We're trying to persuade them to look for more fertile worlds to conquer."

Dorothy sighs. "I'd thank you, but it seems you're trying to take over Earth yourself."

"Maybe not, if you can demonstrate your worth."

"Worth? The worthiness of humanity is not measured by the intelligence of its individuals but rather by the compassion of the race as a whole."

"I'll have to think about that one." Dr. Oz hands Dorothy pieces of pentagonal jewelry. "Take a look at the coelenterates' favorite medallions. They are all shaped like polygons."

"Beautiful."

"But now I want you to consider more generally a regular n-sided polygon, or n-gon. For example, a triangle is a 3-gon and a square is a 4-gon."

"Tell me more. I'm intrigued."

"I want to consider the number of pieces created inside an *n*-gon by its diagonals."

"That's easy. For a square, the diagonals create 4 pieces." Dorothy grabs a chalk and draws a square cut by its diagonals.

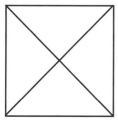

Dorothy cuts a square with its diagonals

"You are correct that the number of pieces is easy to determine for polygons with just a few sides. In fact, we can draw a table of values for triangles, squares, pentagons, and so forth. The variable $R(n)$ is the number of pieces." Dr. Oz holds up his viewscreen for Dorothy to see.

Name	n	R(n)
triangle	3	1
square	4	4
pentagon	5	11
hexagon	6	24
heptagon	7	50
octagon	8	80
nonagon	9	154
decagon	10	220

Polygons cut by their diagonals

"Ooh," Dorothy says, "the number of slices grows fast. There are 220 slices in a 10-sided polygon."

"Okay, Dorothy, here is your challenge. Consider a 30-gon with all its diagonals drawn." The viewscreen shows Figure 58.1. "Your mission is to tell me how many slices this is cut into. Your job is also to derive a formula that will let you count the number of slices for any polygon cut by its diagonals."

58.1. A 30-gon cut to smithereens by its diagonals that have an amazing 16,801 intersection points. (After Bjorn Poonen and Michael Rubinstein.)

"Oh, God no. That is too difficult for me to solve."

Could it be that Dorothy is right and that no human should be expected to solve such a difficult problem?

Difficulty Level: 🜲 🜲 🜲 🜲

59 Cosmic Call

Far-away planets have their plants and animals, and their rational creatures too, and those as great admirers, and as diligent observers of the heavens as ourselves. . . .

> — Christiaan Huygens, seventeenth-century Dutch physicist and astronomer, in *Cosmotheoros, or New Conjectures Concerning the Planetary Worlds, Their Inhabitants and Productions*

Dorothy can see the lines of a satellite dish glimmering like mercury in the noonday sun. All over Kansas these communication devices are sprouting like weeds. It seems as if Dr. Oz wants to turn Kansas into one giant communications station.

"Dorothy, if you could send a message to an alien civilization, what would you say and how would you send it?"

"How about 'Hi, take Dr. Oz away from me so I can return to Aunt Em.'"

"Very funny. Let me tell you something that really happened in 1999. Canadian physicists Yvan Dutil and Stephane Dumas broadcast a custom-designed message in the direction of nearby stars in order to search for extraterrestrial intelligence. The complete message is about 400,000 bits long and transmitted three times over a three-hour period in the direction of four selected stars."

"Is that a lot of bits?"

"This message is much larger in size, duration, and scope than the one sent by Dr. Frank Drake on November 16, 1974, from the Arecibo Observatory. Using a 70-m (230-ft) Ukrainian antenna equipped with a 150-kW transmitter broadcasting at 5 GHz (6 cm), Dutil and Dumas hoped to access civilizations within 100 light-years. Even if more distant civilizations cannot capture all of the information, they should be able to at least discern that it is an artificial signal. The last page of the message invites anyone who reads it to reply."

Dr. Oz tosses a dog biscuit to Toto and then looks back at Dorothy. "I'm leading up to your new puzzle." He pauses. "The Dutil–Dumas message was built to minimize the loss of information due to noise introduced into the signal during its interstellar flight. For maximum accuracy, the special set of characters had significant differences from one another. Redundant information, as well as the presence of a frame around each page, were included to increase the chance of reception and decipherment. At the top of the pages were various symbols to help orient any creature receiving the message."

Dr. Oz hands Dorothy two cards containing symbols (Figures 59.1 and 59.2).

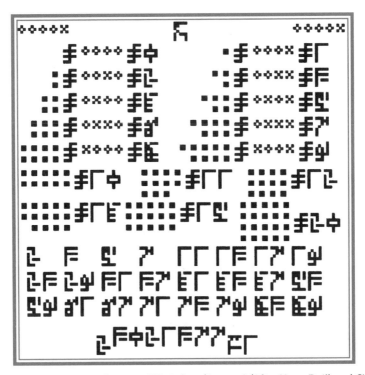

59.1. Message to the stars. What does it mean? (After Yvan Dutil and Stephane Dumas.)

"Dorothy, your mission is to decode at least some of the meaning of the cards. I will give you one month to work on this, at which point you will report your findings to the trochophore council. And they are a pretty mean bunch. If you can figure out the meaning of these messages, I will move our base of operations from your beloved Kansas to New Jersey. Kansas will once again be free."

59.2. Message to the stars. What does it mean? (After Yvan Dutil and Stephane Dumas.)

60 Knight Moves

Many have become chess masters — no one has become the master of chess.
— Siegbert Tarrasch

Dorothy and Dr. Oz are walking along a trail that winds along the creek banks. They pass through native grasslands and into a woodlands area.

Dorothy points to the sky. "Look, a red-tailed hawk."

Dr. Oz nods. "I hear other sounds – wild turkeys, white-tailed deer in the underbrush."

As they approach a forest, the amiable Mr. Plex ambles out between the elms and cedars. He is carrying a sign that says:

> **Trail open sunrise to sundown year-round.**
> **No bikes, roller skates, or skate boards allowed.**
> **All dogs must be leashed.**
> **Please do your part to keep the nature trail clean.**

"Sir," Mr Plex says. "This is a souvenir for my family."

Dr. Oz nods. "Very good Mr. Plex. But pay attention. I have a new puzzle for Dorothy."

Dr. Oz reveals another sign, this one with numbers and shows it to Mr. Plex and Dorothy. "Start with a chess knight on the 2 at top left. Each square you land on – including the first square with a 2 – gets added to your cumulative sum. You may not visit the same square twice. Find a path to Mr. Plex that sums to 18."

2	3	3	2	1	4
1	2	3	7	2	3
3	2	1	1	3	7
1	1	3	2	3	4
2	2	4	3	4	2
7	4	☻	3	2	3

Find a path to Mr. Plex

Dorothy examines the sign. "Do you mean that from the first starting square, I can either hop down two cells and to the right, landing on the 2, or that I can hop two cells to the right and then down one to land on the 3?"

"Yes, but the challenge is to get to Mr. Plex with the sum of 18. You have one hour to solve this. Good luck."

Difficulty Level: ✹ ✹ ✹

61 Sphere

If automobile technology had advanced at the same rate as computer technology, a Rolls-Royce would travel at supersonic speeds and cost less than a dollar.

— William Poundstone, *Labyrinths of Reason*

"**D**orothy, did you know that more than 150 types of grasses and 300 species of wild flowers are in a Kansas tallgrass prairie?" He rips out a few species with his tentacles. "Look, here's some Big Bluestem, Indian Grass, and Switch Grass."

"How can you talk about grasses? I want my freedom."

Suddenly, Mr. Plex descends from the sky and settles a huge spherical spaceship in the prairie:

Spherical spaceship

Dr. Oz motions toward the ship. "The surface area and volume of Mr. Plex's spherical spaceship are both four-digit integers times π, expressed in units of feet. Assume the strange ship contains only air. Because there is nothing else inside, its volume is the usual $(4/3)\pi r^3$. What is the radius r of the alien sphere? If you solve this, I will give you the ship so that you can explore the universe and be free of me."

Difficulty Level: ✸ ✸

62 Potawatomi Target

The computer, insofar as it solves the equations of mathematical physics, insofar as it is an instrument of oracularity and purports to tell us today what will happen tomorrow, collapses time by making the future appear now.

— Philip Davis and Reuben Hersh, *Descartes' Dream*

"**D**orothy, today we are visiting the Potawatomi Indian Mission in Kansas. This building was completed in the spring of 1850 and housed approximately 90 Native American children. The children were taught reading, writing, and basic skills such as needlework and blacksmithing."

"And why are you telling me all this, Dr. Oz?

"Native Americans were experts with the bow and arrow." Dr. Oz hands Dorothy a bow and arrow and points to an unusual target hanging on the wall. "With four shots, hit four different numbers on the target that total 150." Can you do it?

10	111	33	113
43	13	54	12
16	87	93	79
49	89	108	27

Dr. Oz's target

Difficulty Level: 🐜 🐜

63 **Sliders**

The name of Leonardo da Vinci will be invoked by artists to prove that only a great artist can be a great technician. The name of Leonardo da Vinci will be invoked by technicians to prove that only a great technician can be a great artist.

— Alex Gross, *East Village Other*

"**D**orothy, notice that Mr. Plex and his clones have assembled themselves in an array for the next puzzle. Slide three of the aliens up, down, right, or left, so that there are exactly three aliens in each row and column."

Sliders

Dorothy walks through the prairie and begins to examine the peculiar arrangement of Plexes. "What will you give me if I can solve this puzzle?"

"I will give you Cliff Pickover's new book, *Dreaming The Future: The Fantastic Story of Prediction*."

"Excellent choice."

"You have twenty minutes to complete this problem."

Difficulty Level: ✳

64 Swapping

We live under the shadow of a gigantic question mark. Who are we? Where do we come from? Whither are we bound? Slowly, but with persistent courage, we have been pushing this question mark further and further towards that distant line, beyond the horizon, where we hope to find our answer. We have not gone very far.

– Hendrik Willem Van Loon, *The Story of Mankind*

"**O**h, my gosh," Dorothy whispered. "Th – there's hundreds of them."

Dr. Oz and Dorothy are underground looking at sleeping aliens. Each alien appears to be in suspended animation. Their faces are partially covered by some kind of breathing mask with a snaking tube that goes into a hole in the wall. Their fragile limbs are hooked to flashing monitors amid a maze of cables and gauges.

The alien nearest them has skin that is bleached whiter than bone. It stands there against the wall, seemingly lifeless, its insectile head encased in an enormous turban of wires and tape. A thick amber tube runs into its mouth. A thin needle is taped to its right forelimb. The only signs of life are provided by the blinking of the machines – machines that probably keep the creatures alive or quiet in this dormant state.

"Dorothy, we don't know where these aliens come from. But look, one of them has a puzzle in its hands. We believe if you can solve the problem, we will better understand the aliens' purpose."

Dorothy reads the words on the puzzle:

Swap two pairs of numbers so that each row, column, and diagonal sum to the same constant. (As an example, 16 and 7 might be interchanged as one of your two swaps, but this is not part of the right answer.)

16	3	2	13
5	10	11	8
12	6	7	9
4	14	15	1

Swap two pairs

Difficulty Level: 🐜 🐜

65 Triangle Dissection

I looked upon the Moon mission as basically another battle in the cold war. We were fighting the Russians, and we were trying to achieve a mandate from our president. It was not a romantic moment of exploration as such.

— Frank Borman, commander of the *Apollo 8* Moon mission.

Dr. Oz is invited to a conference held by the New York Academy of Sciences in New York City. The conference is funded by the SETI Institute, a group devoted to the search for extraterrestrial intelligence. In the audience are scientists, UFO buffs, and TV and movie celebrities.

Dr. Oz masks his squidlike appearance so that the attendees believe him to be an Armani-suited human traveling with his daughter Dorothy and her dog. His audience grows hushed when he shows them a triangle with one obtuse angle (an angle greater than 90 degrees) (Figure 65.1). The obtuse angle is at the top of the triangle.

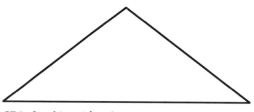

65.1. An obtuse triangle

"My fellow Americans. Here is a test I would give to extraterrestrials to assess their intelligence. Can you cut this triangle into smaller triangles, all of them acute? (An acute triangle is a triangle with three acute angles; that is, angles less than 90 degrees)."

Lucy Liu, Drew Barrymore, and Cameron Diaz – stars of the movie *Charlie's Angels* – step closer to a microphone. Lucy says, "Sir, can the triangles contain a right angle?"

"No, a right angle is neither acute nor obtuse."

Drew and Cameron first try with the following dissection, sketching with an electronic pen on a Web-browser window, but the solution leads nowhere (Figure 65.2). Cameron laments, "Skaffa ny bläddrare för jävulen!"

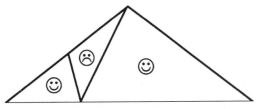

65.1. Charlie's Angels' first attempt at dissection

Dr. Oz draws smiley faces on Cameron's acute triangles and a sad face on the obtuse one. "Cameron, good try. Two of your triangles are acute, which is good, but if you carefully look at the third triangle, you'll see that it is obtuse. Can you divide the triangle to produce all acute internal triangles?"

He has not noticed that Toto has leapt from Dorothy's arms and clamped his jaws around the hem of the cloaking device. With one swift yank, Dr. Oz is revealed in all his trocophorean glory. Scientists who'd long hoped to contact alien species find themselves unprepared for so sudden a confrontation and rapidly retreat to plan their next move – for some, perhaps a good stiff drink.

In the pandemonium, Dr. Oz spots a young Kansan pressing toward the door. A well-aimed tentacle pulls Dorothy and Toto off the floor, reels them in, and binds them tightly to him.

Difficulty Level: ✺ ✺ ✺ ✺

66 A Simple Code

Mathematical inquiry lifts the human mind into closer proximity with the divine than is attainable through any other medium.

— Hermann Weyl (1885–1955)

As the humans on Earth start to realize that they are sharing their planet with Dr. Oz and his trochophore kin, many human scientists begin to have difficulties finding funding for their research. One side effect of the arrival of Dr. Oz is that governments and politicians assume Dr. Oz will give humanity access to his advanced knowledge. Why should Earthly scientists spend valuable dollars researching areas that the trochophores have already fully mined?

Dr. Oz tells Earthly scientists that he will fund their research – and let them study several trochophore patents for advanced technology – if Earthlings can find anyone who can solve this puzzle within two minutes.

What number replaces the question mark?

2	0	0	0	0	1
1	4	0	0	1	0
2	0	1	1	5	0
3	1	0	3	2	3
0	1	1	0	?	0

Difficulty Level: ✷ ✷

67 Heterosquares

More significant mathematical work has been done in the latter-half of this century than in all previous centuries combined.

— John Casti, *Five Golden Rules*

"**D**r. Oz, everyone on Earth knows about your existence. There's nowhere to run, nowhere you can hide. You can't keep me prisoner much longer."

Dr. Oz nods. "I know, Dorothy. How about I set you free within 14 days? You have proved yourself to be very intelligent."

"Sounds good." Dorothy says. "As you know, the United Nations is currently debating ways they can manage the public response to the news of your species cohabiting Earth with us. The UN has even paid teenagers money to say calming things about you in Web chats, news groups, and on MTV. I guess the government doesn't want a panic."

"OK, let's get back to work. Here is your puzzle for today. Just because I will set you free doesn't mean I will stop assessing your intelligence." He pauses. "I have given this puzzle to all teenagers watching MTV this week. Place the consecutive numbers 1 through 9, one number in each cell, so that the rows, columns, and main diagonals have different sums."

Dr. Oz hands Dorothy a card with an empty grid.

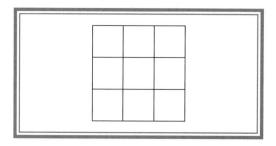

She traces her fingers over the grid. "This problem reminds me of magic squares, where the numbers in every row, column, and diagonal produce the same sum. But here we have to find the opposite."

Dr. Oz nods. "Perceptive. Now get to work!"

Difficulty Level: ✹ ✹

68 **Insertion**

An intelligent observer seeing mathematicians at work might conclude that they are devotees of exotic sects, pursuers of esoteric keys to the universe.
– Philip Davis and Reuben Hersh, *The Mathematical Experience*

Dr. Oz and Dorothy walk through a maze of stalagmites and moist passages that wind through one another like worm tunnels.

"Watch out!" Dr. Oz screams.

Dorothy is slipping feet first into a funnel-like pit that drops off an indeterminable distance. Dr. Oz reaches out his tentacle to grab her.

"No," Dorothy cries, "I'll take you with me."

"No you won't," Dr. Oz yells. "Now grab my tentacle." His voice echoes around the cavern like the wails of ghosts.

A sharp crystal rips though Dr. Oz's newly retailored Armani suit as he spreads his tentacles in an attempt to stop Dorothy's descent.

"I'm OK," Dorothy says holding his tentacle. Slowly she manages to climb her way out of the pit. Her body shakes. "That was close," she says.

"Dorothy, don't think that I'm getting emotionally weak. I was not acting purely out of compassion when I saved you. I wanted you alive so I could test you with my next puzzle."

"I don't believe you!"

"Believe this." Dr. Oz pulls out a card and hands it to Dorothy. "Insert +, −, ×, ÷ and/or parentheses between each digit to find the total."

$$8\ 7\ 6\ 5\ 4\ 3\ 2\ 1 = 36$$

Dr. Oz continues. "For example, one could say $(87 - (65 \times 4)) \div 321$, which, of course, is not correct because it does not yield 36. You have one hour to solve the problem.

Difficulty Level: ✳

69 Missing Landscape

There is no question about there being design in the Universe. The question is whether this design is imposed from the Outside or whether it is inherent in the physical laws governing the Universe. The next question is, of course, who or what made these physical laws?

— Ralph Estling, *The Skeptical Inquirer*

Dorothy and Dr. Oz continue their journey across the Midwest from Kansas to Oklahoma to Arkansas. They travel light these days: just toothbrushes, a couple of tunics made from a self-cleaning material, and several sets of undergarments. A flexscreen is all Dr. Oz needs to continue testing Dorothy and to connect them to the ever-shifting reality of an America cohabited by humans and trochophores.

"Dorothy, here is your next test." He shows her an array of landscapes on the viewscreen. What is the missing landscape?

Landscape shatter

Difficulty Level: ✳ ✳ ✳

70 **The Choice**

To those who do not know mathematics it is difficult to get across a real feeling as to the beauty, the deepest beauty, of nature. . . . If you want to learn about nature, to appreciate nature, it is necessary to understand the language that she speaks in.

— Richard Feynman, *The Character of Physical Law*

"Dorothy," Dr. Oz says, "I want you to visit ten Internet Web pages, totally at random. Each page has a certain number of words. For example, some pages might only have 15 words. Others might have thousands. As you browse each site, your aim is to stop browsing when you come to a Web page that you guess will have the largest number of words. You cannot go back and pick a previous page you browsed. If you browse all ten pages, then, of course, you must pick the last page browsed. Once you have selected a page, you'll continue to browse randomly until all ten pages have been displayed. Let's assume each Web page has a different number of words."

Dorothy first chooses a Web site at random using a random character generator. She happens to select www.pickover.com, which has 1,000 words.

What do you think are her odds of selecting the page with the highest number of words? What is her best strategy? When should she stop browsing and say, "This is the one, Dr. Oz"? If you follow her best strategy, how can you calculate your chances of winning?

Difficulty Level: ✸ ✸ ✸ ✸

71 Animal Selection

Some people think of God . . . busily tallying the fall of every sparrow. Others — for example, Baruch Spinoza and Albert Einstein — considered God to be essentially the sum total of the physical laws which describe the universe.

— Carl Sagan, *Broca's Brain*

"Welcome to the University of Kansas," Dr. Oz says.

Dorothy looks all around at the red and orange buildings with huge numbers on their sides. "It's not quite the same as I remember it."

The buildings are scattered over miles of grassy plain. The university itself seems to be surrounded by an unending array of fast-food places, which, for some odd reason, are crammed full of mammals of the nonhuman persuasion. Many are elephants.

"Dorothy, we are changing your world into something new and beautiful. Do not worry. You'll grow to like it. All these animals inspired me to create a new puzzle for you." He brings out a flexscreen displaying eight patterns, each containing three animals:

"Your mission is to replace the question mark at the bottom right with one of the following nine patterns."

Dr. Oz steps very close to Dorothy. "Which one of the nine patterns would you use to complete the initial set and why?"

Difficulty Level: ✳ ✳

72 The Skeletal Men of Uranus

If we wish to understand the nature of the Universe we have an inner hidden advantage: we are ourselves little portions of the universe and so carry the answer within us.

— Jacques Boivin, *The Single Heart Field Theory*

"**D**orothy, I'd like to introduce you to the skeletal men of Uranus. They spend their lives dancing to melodies we can never hear."

Dorothy looks at the dancing men. Their living quarters are air-conditioned to a chilling 40 degrees F. She shivers as she hold Toto tightly in her arms.

"My colleagues have arranged themselves into eight patterns, each containing one or more men."

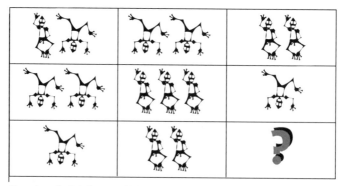

Dancing skeletal men of Uranus

"In the bottom right cell, I have erected a metallic question mark symbol. Replace the question mark with one of the three sets of men that stand before you."

Which of the three sets should Dorothy choose?

73 Hindbrain Stimulation

The hidden harmony is better than the obvious one.

— Heracleitus, c. 540–c. 480 B.C.

Dorothy is strapped to a chair. Dr. Oz is tickling her hindbrain with an electrical probe so that a series of images appear to float in her visual field.

"Dorothy, by stimulating your visual cortex, I can make you see all sorts of patterns. Do you see the array of hands? For this puzzle, you have to supply the missing hand."

Symbols flashing before Dorothy's hindbrain

Difficulty Level: 🕸🕸

74 Arrays of Absolution

Truly the gods have not from the beginning revealed all things to mortals, but by long seeking, mortals make progress in discovery.

— Xenophanes of Colophon, c. 560–c. 478 B.C.

Dorothy gazes at an obelisk perhaps 12 feet high and wrought in gold. A tangle of yellow tendrils trail to the ground from several orifices. Along the surface of the obelisk is a series of arrays:

1	0		0	1		2	2		1	2		4	2
2	1		1	3		0	2		2	2		0	2

Arrays of Absolution

"Dorothy," says Dr. Oz. We call these the Arrays of Absolution because any trochophore who can determine which of the following arrays completes the series is absolved of all sin."

2	2		3	1		2	2		3	2		0	0
1	3		2	3		1	2		0	2		0	0

Choose one array to complete the series

"Dorothy, choose one array."

Difficulty Level: ✸ ✸

75　Trochophore Abduction

The pure mathematician, like the musician, is a free creator of his world of ordered beauty.

　　　　　　　　　　　　– Bertrand Russell, *A History of Western Philosophy*

Dr. Oz is abducting Earth animals for an intergalactic zoo. Inside Dr. Oz's ship in this round are five pairs of bison, four pairs of elephants, and two pairs of zebras. (A pair consists of a male and female.)

Bison, elephants, and zebras

When Dr. Oz reaches a huge ark in outer space, he starts to open a chute that lets animals drop from the ship, one at a time, into individual cages. Later, Dr. Oz wants to match the species, and pairs within a species. The lighting system is malfunctioning – some giraffes, before being caged themselves, had nibbled at the peripheral glow-spheres – and there is insufficient starlight for Dr. Oz to differentiate between the different new animals visually. How many of the new arrivals must Dr. Oz drop to ensure that he has two animals of the same species? How many animals must he drop to ensure that he has a male and female of the same species?

Difficulty Level: 🦌

76 The Dream Pyramids of Missouri

What could be more beautiful than a deep, satisfying relation between whole numbers. How high they rank, in the realms of pure thought and aesthetics, above their lesser brethren: the real and complex numbers. . . .

— Manfred Schroeder, *Number Theory in Science and Communication*

Dorothy is with Dr. Oz in Missouri. For as far as the eye can see are pyramids made of glass, copper, and steel. The thousands of pyramids slowly rotate at different rates and make an eerie, crying sound, like wind rushing through a deep cave.

"Dr. Oz, what have you done to Missouri?"

"The pyramids are a homage to one of my favorite puzzles. Notice how each pyramid has the same series of numbers on them. Trochophore philosophers sometimes dream within the pyramids in order to solve a centuries-old problem."

Dorothy steps closer to a pyramid to observe the numbers.

1 1 2 3 3 4 4 5 7
**Pattern of numbers on the
dream pyramids of Missouri**

"Dorothy, using all the digits, you are to form a single fraction that equals one-eleventh."

"Do you mean like 33211/5744?"

"Yes, but you can see that your example fraction is not equal to $\frac{1}{11}$. You have five hours to form the correct fraction or else we will plant a similar array of pyramids in Mississippi, Kansas, and Alabama – something your fellow Earthlings would probably prefer not happen."

Difficulty Level: ✴ ✴

77 Mathematical Flower Petal

I do not know what, if anything, the Universe has in its mind, but I am quite, quite sure that, whatever it has in its mind, it is not at all like what we have in ours. And, considering what most of us have in ours, it is just as well.

— Ralph Estling, *Skeptical Inquirer*

"**W**ow," Dorothy says, "That's awesome." She gazes at a huge flowerlike edifice that has sprouted from the swamps near New Orleans, Louisiana. The following pattern is engraved on each petal:

	×	2	+		=	
+		+		×		+
		10				
−		+		+		+
	+		×	3	=	
=		=		=		=
	+		+	19	=	

Petal of the Louisiana flower

"I knew you would appreciate it. Every day the flower opens in the morning and closes at night. Trochophores climb inside it and must insert numbers into the empty cells so that all the mathematical expressions are correct, reading from left to right and top to bottom. If they fail to solve the puzzle by nightfall, the petals close and trap them in a deadly embrace. Dorothy, can you solve the puzzle?"

Difficulty Level: ✹ ✹

78 Blood and Water

The belief that the underlying order of the world can be expressed in mathematical form lies at the very heart of science. So deep does this belief run that a branch of science is considered not to be properly understood until it can be cast in mathematics.

— Paul Davies, *The Mind of God*

Dr. Oz places two golden goblets on a table. "Dorothy, one goblet contains blood, the other water. Both goblets contain exactly the same volume of liquid."

Blood   Water

Dr. Oz takes a teaspoon of water from the water-filled goblet and mixes it into the blood-containing goblet and then takes a teaspoonful from the blood goblet and mixes it with water. Both goblets are now contaminated.

"If you can determine which goblet is more contaminated, you will be set free. Does the water now contain more blood than the blood does water, or the other way around?"

Difficulty Level: ✷ ✷

79 Cavern Problem

If you were a research biologist from the Stone Age, and you had a perfect map of DNA, could you have used it to predict the rise of civilization? Would you have foreseen Mozart, Einstein, the Parthenon, the New Testament?

— Deepak Chopra, *Ageless Body, Timeless Mind*

Dr. Oz and Dorothy are deep inside a cave in western Tennessee. Entering quietly, they stand still and survey a bizarre, cold world of emaciated sleepers stacked against the cavern walls.

"Oh, my God," Dorothy whispers. "Th – there's dozens of poodles."

Each poodle appears as if it is in suspended animation. Their muzzles are partially covered by aluminum foil. Their hindlimbs make contact with large metallic plates.

Dr. Oz nods. "They are poodle–human hybrids. We are raising them as a hive unit and will focus their minds on one and only one task, which our philosophers have been refining for decades as a test of interspecific bimentalism. Take a look at the problem." Dr. Oz points to a poster on the wall.

4	2	28	15	89
1	3	40	70	35
8	7	18	30	46
5	6	48	27	42

"One of the gray cells should be white, and one of the white cells should be gray in order to fit a hidden numerical relationship. Which cells should you change?"

80 Three Triplets

The future is a fabric of interlacing possibilities, some of which gradually become probabilities, and a few which become inevitabilities, but there are surprises sewn into the warp and the woof, which can tear it apart.

— Anne Rice, *The Witching Hour*

"**A**nd what are these five poodles doing?" Dorothy asks.

The poodle nearest Dorothy and Dr. Oz is hooked to some kind of brain-monitoring device. The animal reclines on an oak platform, nearly lifeless. The only sign of animation is the incessant tapping of the poodle's left foot on a large computer keyboard.

"These genetically enchanted animals are contemplating the following problem. Consider the three triplets of symbols on the wall. Note that the set of bones is to be thought of as a single symbol."

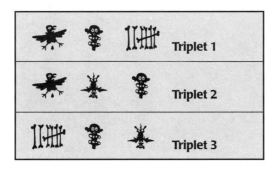

"Dorothy, the poodles must choose one of the following that best accompanies the above group."

"Can you help the poodles find a solution and give a good justification for your answer?"

81 The Oos and Oob Gambit

Science fiction, like science, is an organized system that, for many, takes the place of religion in the modern world by attempting a complete explanation of the universe. It asks the questions – Where did we come from? Why are we here? Where do we go from here? – that religions exist to answer. That is why *religious* science fiction is a contradiction in terms although *science-fiction* about *religion* is commonplace.

– James Gunn, *The New Encyclopedia of Science Fiction*

Dorothy is with Dr. Oz and Mr. Plex. To their left, lying in lucite incubators lit by fluorescent lights, are small creatures, only a foot or two in length, perhaps fetuses – living breathing things, eyes still sealed shut. Their spindly arms and gross, oversized heads remind Dorothy of human babies with medical deformities. Why are they being kept alive like this? Are they just the first step in the creation of the adult?

"Dorothy, these aliens are contemplating a very difficult problem. Let me show you."

Dr. Oz places Dorothy in a large jar with two kinds of hideous, squirming creatures, the Oos and the Oob.

Oos   Oob

"Dorothy, I will place one Oos and one Oob into an opaque flask. You will have to withdraw a creature from the flask without looking.

If the creature is an Oos, your dog Toto will be placed in a state of suspended animation, and you will not see him for thirty years. If it is an Oob, I will set you free. These creatures are fast. Don't drop one or it will scurry away so quickly that we won't be able to see it or recover it."

"Dr. Oz, why do you place me in such outrageous situations? I know you want to test me, but this is getting a bit ridiculous."

"Do you dare challenge the great and powerful Oz?" Dr. Oz bends down and picks up two creatures to put in the flask, but Dorothy notices that he has cheated and put in two Ooses. She wants to scream out, "Cheater!" However, this may cause Dr. Oz to lose face in front of the honest Mr. Plex. Oz's resulting anger could be dangerous for Toto. How can Dorothy seem to go along with the plan and save Toto, now that she knows that there are two Ooses in the flask?

Difficulty Level: 🗲

82 Napiform Mathematics

The brain, knowing that a person can't live forever, rationalizes a future, other-dimensional world in which immortality is possible.

— Philip José Farmer

Dorothy gazes in horror as the small stalklike arms of alien fetuses begin to curl as if magically coming to life. "Why are you growing so many of these creatures?" she whispers as hers eyes dart from one incubator to the next. The fetuses' small, bony mouths seem to open in a silent cry. A few napiform heads gasp for breath.

"These beings are working on yet another number problem. If you can solve the problem before they do, you get one million dollars. You must arrange the consecutive numbers 1 through 16 (one number to a cell) such that each row, each column and each of the two long diagonals produces different sums, and the sums form a consecutive series of integers.

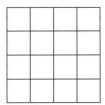

Fill the array

Difficulty Level: ✳ ✳ ✳

83 Toto, Mr. Plex, Elephant

Theology is a branch of physics. . . . Physicists can infer by calculation the existence of God and the likelihood of the resurrection of the dead to eternal life in exactly the same way as physicists calculate the properties of the electron.

— Frank Tipler, *The Physics of Immortality*

Dr. Oz, Mr. Plex, Toto, and Dorothy walk through Dr. Oz's art gallery. De Kooning paintings are hung on the walls. All the paintings have hooked, recurving lines through dense paint. Abstract expressionistic designs also cover some of the stalagmites protruding from the museum floor.

"Dr. Oz, you have unusual tastes."

"I'll take that as a compliment. Follow me."

Dr. Oz points to a particularly enigmatic painting filled with portraits of Mr. Plex, Toto, and elephants:

🐘🐕	🐕🕷	🕷🐕	🐘🐕
🕷🐕	🐕🕷	🕷🐕	🐕🕷
🕷🐕	🐘🐕	🐘🐕	🕷🐕

An enigmatic painting

"Dorothy, you are to start at one cell and create a continuous path through the entire array that goes though every cell. You move right, left, up, or down, and your path may not intersect itself. No two consecutive animal pairs may be the same. For example, you cannot go from a cell with 🦁🐎 to an adjacent cell with 🦁🐎, although you can go to an adjacent cell with the opposite ordering 🐎🦁 or to a completely different set such as 🐘🐎."

Difficulty Level: 🦋

84 Witch Overdrive

A friend of mine once was so struck by [a recursive plot's] infinitely many infinities that he called it "a picture of God," which I don't think is blasphemous at all.

— Douglas Hofstadter, *Goedel Escher Bach*

Dorothy and Dr. Oz look up to the sky and watch as a wicked witch draws patterns with the exhaust from her broomstick. Two of the patterns are shown in Figures 84.1 and 84.2.

"Dorothy, what you see is an android witch flying on a novel propulsion device."

"She must get awfully dizzy."

"I admit the patterns are complex, but they are based on Lissajous figures generated by the following C program."

"What's a 'C program'?"

"Never mind that. Just think of it as a simple recipe." Dr. Oz hands her a card.

```
For (t=0.0; !isclosed(); t+=Tstep) {
    x= Xamplitude * sin(Xfrequency * t +Xphase);
    y = Yamplitude * sin(Yfrequency * t+Y phase);
    Plot (x,y);
}
```

Wicked witch C code

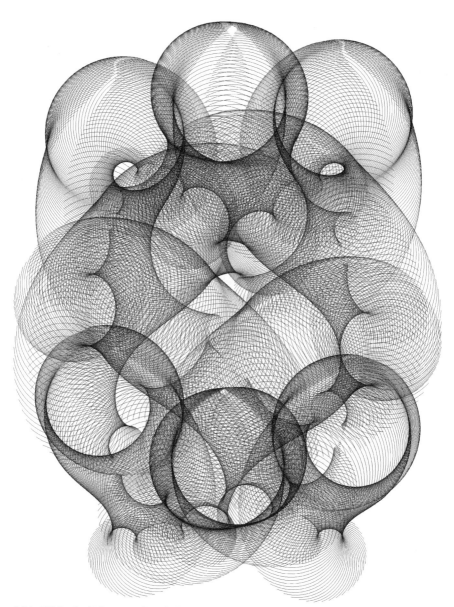

84.1. Wicked witch curve. (Rendering "Concerto for Horns" by Bob Brill.)

"All you have to do is supply values for parameters – the capitalized variables in the computer recipe. As the value of t increases, the resultant values of x and y trace out an intricate path. Luckily for the witch, Lissajous figures are well behaved. For example, they are continuous at every point and form sweeping curves that suggest their

Witch Overdrive 177

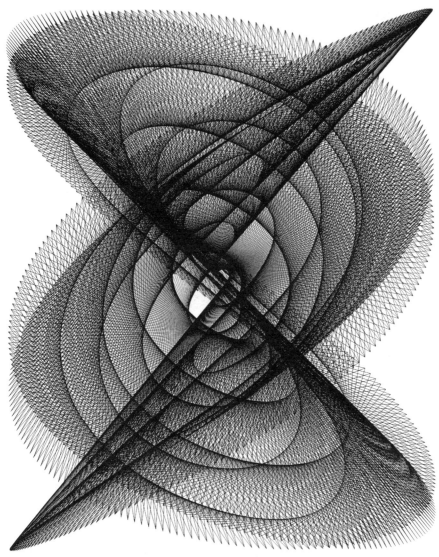

84.2. Wicked witch curve. (Rendering "Lissajous Pursuit 2" by Bob Brill.)

sinusoidal function origins. Do you think such curves always return to their starting points?"

"Not sure."

"If you like, you can create an 'is-closed function' that returns a value of 'true' when the current x,y position is the same as the starting coordinates or when the curve appears to repeat on top of itself."

"The images are so beautiful. There must be more to their generation."

"Yes, there is a bit more. Can you figure out what extra embellishments are needed to create the figures? Also, can you tell me the maximum and minimum values that x and y will ever have?"

Difficulty Level: ✳ ✳

85 What Is Art?

I keep the telephone of my mind open to peace, harmony, health, love, and abundance. Then, whenever doubt, anxiety, or fear try to call me, they will keep getting a busy signal, and soon they'll forget my number.

— Edith Armstrong

Dorothy wanders around the Oz testing facility. She tries to relax, letting all the anxieties of the last few months leave her mind like clouds fleeing in an autumnal breeze.

She looks all around the strange gallery of forms and sees a De Kooning oil painting. Underneath the painting is a tag giving the title of the painting: *Woman I,* 1950. The huge pupils of the woman in the painting seemed to glare at Dorothy like black headlamps. The woman looks like an opera singer from outer space, and she has the worst overbite in all of Western art.

"Like the painting?" Oz says.

"Not my cup of tea."

"How about this?" Dr. Oz points to a painting of numbers.

327891	*327855*	*327864*
327849	*327828*	*327837*

"You call that art?

"Yes, I do. Which one of these six numbers does not belong? You have six days to give me an answer."

Difficulty Level: ✸ ✸

86 Wendy Magic Square

The peculiar interest of magic squares lies in the fact that they possess the charm of mystery. They appear to betray some hidden intelligence which by a preconceived plan produces the impression of intentional design, a phenomenon which finds its close analogue in nature.

— Paul Carus, in W. S. Andrews, *Magic Squares and Cubes*

Dorothy and Dr. Oz walk through the bowels of the Oz testing facility. Despite Dr. Oz's best efforts, the facility is messy and cluttered, like an overused attic. Electronic equipment is everywhere, stuck to the walls and floors with cable, Velcro, and ropes. If Dr. Oz is not careful, he could cause an avalanche of Pentium chips, CDs, recording equipment, petri dishes, electromyography devices, and electroshock machines.

"Dr. Oz, we've been together for weeks in your Oz testing facility. Don't you have any shower facilities for me?"

"Cleanliness is no problem." Dr. Oz snaps his tentacles together and a robot spider named Wendy scurries along some of the overhead pipes and then leaps into Dorothy's hair. In seconds, Dorothy's hair is cleaned and coiffed into a magnificent hairdo.

Dorothy opens her eyes wide. "Wow, I guess that's a bit better."

"I'm glad you are more comfortable. This should put you in the mood for the next problem." Dr. Oz hands Dorothy a card.

🕷	199	🕷
619	1039	1459
🕷	1879	🕷

Fill in the missing numbers

"Dorothy, fill in the missing numbers – which Wendy is indicating along with her spider friends – so that each row, column, and diagonal sum to the same number. When you solve the problem, tell me what's remarkable about the numbers and this magic square."

Difficulty Level: 🕷 🕷

87 Heaven and Hell

In the pure mathematics we contemplate absolute truths which existed in the divine mind before the morning stars sang together, and which will continue to exist there when the last of their radiant host shall have fallen from heaven.

— Edward Everett, quoted by E. T. Bell in *The Queen of the Sciences*

Dr. Oz and Dorothy float in a craft about 100 feet across. The vessel is shaped like an oval, at the center of which is a squidlike robot sprouting seventeen tentacles that articulate as they amble back and forth across wet surfaces. Dorothy cannot identify the robot's purpose. The robot begins to chant in a low voice.

"Dorothy," says the robot, "your concepts of heaven, hell, and limbo are interesting to my species."

Dorothy gasps and then responds to the robot. "I like to think of them as states of mind rather than actual places."

"Nonsense," says the robot. "I have visited hell in my dreams. It is ruled by a large squid with sharp tentacles."

Dr. Oz nods. "What the robot says is true," he says.

Dorothy shakes her head. "Dr. Oz, don't you see that you are projecting your own hopes, fears, and images onto an outmoded concept of the afterlife?"

"Dorothy," Dr. Oz says, "you are beginning to speak in a manner too sophisticated for a Kansas farm girl, even at the turn of the twenty-first century."

"Yes, I am aware of that. Mr. Plex massaged my medula oblongata and cerebrum, thus enhancing their ability to process more complex information."

Dr. Oz stamps his tentacle on the floor. "Well, process this." He hands Dorothy an engraved golden plaque.

	heaven	hell	heaven	
heaven	limbo	limbo	limbo	hell
hell	hell	heaven	hell	heaven
limbo				limbo
heaven				hell
hell				heaven
hell	heaven	hell	limbo	limbo
limbo	heaven	heaven	heaven	hell
	hell	limbo	hell	

Trochophore vision of life and afterlife

"What's the longest path you can take through this maze?" Dr Oz asks. "You can start and end on any cell, but you cannot go through a cell more than once. You move up, down, right, and left and may not repeat adjacent words as you travel. For example, your path cannot travel through 'limbo limbo.'"

88 The Stars of Heaven

With equal passion I have sought knowledge. I have wished to understand the hearts of men. I have wished to know why the stars shine. And I have tried to apprehend the Pythagorean power by which number holds sway about the flux. A little of this, but not much, I have achieved.

— Bertrand Russell, *The Autobiography of Bertrand Russell*

Dorothy is stroking a squid the size of an apple. With each stroke, the baby trochophore purrs. Puffs of black smoke and little sprays of crystals spring from several orifices.

Dr. Oz nods in apparent appreciation. "Dorothy, thank you for taking care of my aunt's baby."

The crystal effluence glitters like diamonds in the noonday sun. "What's her name?"

The baby's tentacles writhe like a nest of snakes. "We call her Emma. Emma's aunt gave me a new puzzle for you."

Dr. Oz hands Dorothy a flexscreen with the following configuration of stars.

	1	2	3	4	5	6
1		★	★	★		
2	★	★	★	★	★	
3		★		★	★	★
4	★	★	★	★	★	★
5		★	★	★		
6			★	★	★	

Stars of heaven

"By connecting the centers of stars, it is possible to form various squares. For example, you can see a square indicated by four stars that I have drawn larger then the rest. These stars are at coordinates (2,5), (3,6), (4,5), and (3,4). To make the problem more difficult, you are to find a set of squares such that no two squares share a corner. Can you draw all such squares so that each star is accounted for?"

"Accounted for?"

"Yes, every star must be on the vertex of a square. Given all these rules, show me all your squares. The sample square I just showed you does not have to be part of your solution."

89 Vacation in the Tarantula Nebula

The mathematical life of a mathematician is short. Work rarely improves after the age of twenty-five or thirty. If little has been accomplished by then, little will ever be accomplished.

— Alfred Adler, "Mathematics and Creativity"

Dorothy and Dr. Oz shoot toward the heart of the glorious Tarantula Nebula. Their spaceship's wings are half furled as the ship slowly rotates on its long axis. Robot spiders swarm over the ship's hull making repairs and searching for any weakness. A few robots approach Dorothy while clenching metallic numerals: 1, 2, and 3.

"Dorothy, I want you to meditate on the number 123. Now complete the following sequence."

023   032   113   131   212   ?

"Ooh, this is a beauty. Very difficult. How much time do I have?"

"Five days, three hours, and two minutes. What is the missing number? You may radio the problem back to friends on Earth using my new tachyon-wormhole radio, if this is of help. If you get this problem wrong, I won't take you on the tour of the Tarantula Nebula."

Difficulty Level: 🕷

90   Hot Lava

"**D**orothy, plunging an arm into hot lava is dangerous, with a 50% chance of survival once you remove the remains of your arm."

"This sounds sick."

"No, we're only doing a thought experiment. Imagine this. Uncle Henry, Aunt Em, and Dorothy plunge their arms into lava in succession. For example, first Uncle Henry is to plunge and remove his arm. Next Aunt Em. Finally, Dorothy. The *winner* will be the *first* to survive. The game stops when someone wins or after Dorothy plunges her hand into the lava, whichever comes first. In other words, each person has only one shot at winning or losing. What are their respective chances of winning?"

"You are one sick trochophore!"

"I may be sick, but your brain is improving with every problem I ask. Now tell me the odds!"

Difficulty Level: ✺

91 Circular Primes

Computers are composed of nothing more than logic gates stretched out to the horizon in a vast numerical irrigation system.

— Stan Augarten, *State of the Art: A Photographic History of the Integrated Circuit*

"**W**hat's that?" Dorothy asks as she stares at a fat, fetid blob of gelatinous putrescence. The horrid mass is a couple of feet across.

Dr. Oz gives it a poke and watches its shimmering surfaces. A tangle of limbs seethed within its wet surfaces, and occasionally a large eye winks at them. "I have no idea. Let's watch it as we discuss the next problem. As I told you before, a prime is a positive integer that cannot be written as the product of two or more smaller integers. For example, 11 cannot be written as a product of two or more smaller factors; therefore, 11 is called a prime number or prime."

Dorothy eyes the gelatinous mass. "Yes, I know all about prime numbers."

"Well, here's something you don't know anything about: *circular primes*. These are primes with a truly wonderful property. One can find them by repeatedly chopping away the leftmost digit and appending it to the other end of the number. For circular primes, you can repeat this process and finally return to the starting number. If all the intermediate numbers formed in the chopping process are prime, then the starting number can be called 'circular.'"

"Give me an example."

"Certainly, Dorothy." Dr. Oz draws a chain of circular primes starting and ending with 1193.

$$1193 \Rightarrow 1931$$
$$\Uparrow \qquad \Downarrow$$
$$3119 \Leftarrow 9311$$

"1193, 1931, 9311, and 3119 are all primes!"

"Stupendous!"

"Your mission is to find one other circular prime. But don't even think about giving me a trivial answer with a one- or two-digit prime. Can you find another circular prime?"

"I'll do anything if you take me away from the blob."

"I will do so. But keep this in mind. You have one hour to solve the problem, or else we will revisit the nearly liquid mass of loathsome lymph."

Difficulty Level: ✸ ✸

92 The Truth about Cats and Dogs

Throughout the 1960s and 1970s devoted Beckett readers greeted each successively shorter volume from the master with a mixture of awe and apprehensiveness; it was like watching a great mathematician wielding an infinitesimal calculus, his equations approaching nearer and still nearer to the null point.

— John Banville, *The New York Review of Books*

After several months with Dr. Oz, Dorothy has stopped yearning to return to Aunt Em and Uncle Henry and her daily chores of feeding the hogs and chickens. She has discovered that, for the most part, she likes Dr. Oz's puzzles. She likes the strange sounds he makes, the music of the prairies beneath his tentacles, his strange assortment of friends. Sometimes Dorothy spends her time reading a little, drinking Mountain Dew, studying puzzles, and yearning to exploring space–time to degrees she could have barely imagined on her Kansas farm.

Today Dorothy is with Dr. Oz, leisurely strolling through a seemingly limitless plain of wheat. "Dorothy, imagine an array of robot cats and dogs. Six dogs – two of which are shown here – are trying to hunt the cats."

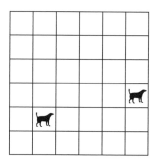

Two of six dogs seeking cats

"Dr. Oz, thank you for not trying to clone Toto for this puzzle."

"You are welcome. Now listen up. The clever cats have found a way to hide from the dogs, provided that the dogs are in certain cells. From its cell, a robot dog can see in all straight-line directions (horizontally, vertically, and diagonally). Can you add four more dogs and three cats in such a way that none of the dogs can see any of the cats? No more than one animal is allowed to occupy a cell."

Difficulty Level: ✺ ✺

93 Disc Mania

For a physicist, mathematics is not just a tool by means of which phenomena can be calculated, it is the main source of concepts and principles by means of which new theories can be created.

— Freeman Dyson, *Mathematics in the Physical Sciences*

"**D**r. Oz, your esophageal sphincter is showing."

"Thank you Dorothy, but that it is quite normal for us to evaginate our digestive system through our oral cavity in times of great emotion. Today I'm so emotional because of a problem involving small discs. I just swallowed one by accident."

"I don't want to hear about your digestive misadventures. Tell me about the puzzle."

"Certainly. Take a look at this." Dr. Oz displays an array of brightly colored discs on a flexscreen embedded in one of his larger tentacles:

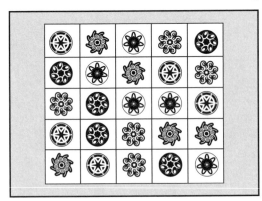

"Switch the positions of three pairs of discs so that each row and column contains five different kinds of discs."

94 $n^2 + m^2 = s$

The bottom line for mathematicians is that the architecture has to be right. In all the mathematics that I did, the essential point was to find the right architecture. It's like building a bridge. Once the main lines of the structure are right, then the details miraculously fit. The problem is the overall design.

— Freeman Dyson, *The College Mathematics Journal*

Dorothy and Dr. Oz are strolling down Main Street in Wichita, Kansas. Some of the aliens appear to have purchased an array of clothing, furniture, and farm equipment, each with a large tag that says:

Proudly Made in the USA by Union Labor

A few of the more thrifty aliens purchased the same items at Dr. Oz's store for half the price. Their tags simply said:

Made by Aliens

It remains to be seen just how the alien competition would affect the local economies.

One shiny tag on a tractor particularly catches Dorothy's eye. It reads:

$n^2 + m^2 = s$

"Dr. Oz, what's that all about?"

"That's one of our most fun number-theory problems. Dorothy, on average, what is the number of ways of expressing a positive integer s as a sum of two integral squares, $n^2 + m^2 = s$?"

Dorothy stares at Dr. Oz, and he grins. Several of his detachable teeth drop to floor and make a crackling sound, like twigs burning in a fire. Perhaps Dr. Oz enjoys Dorothy's confusion.

"I don't follow you," Dorothy says. "Can you say that again?"

"Yes, I would be happy to clarify. On average, over a large collection of integers, how many ways can an integer be written as the sum of two square numbers? For example, 2 can be created four ways." Dr. Oz sketches on a blackboard.

$$1^2 + 1^2 = 2$$
$$-1^2 + -1^2 = 2$$
$$1^2 + -1^2 = 2$$
$$-1^2 + 1^2 = 2$$

"And the number 3 can't be created at all using integers. I repeat my question. On average, what is the number of ways of expressing a positive integer s as a sum of two integral squares, $n^2 + m^2 = s$?"

"Dr. Oz, what do you mean here by 'average'?"

"I want you to add up the number of representations for all numbers up to N, divide by N, and let N get bigger and bigger. For example, because the number of representations of $0, 1, 2, \ldots, 10$ is 1, 4, 4, 0, 4, 8, 0, 0, 4, 4, and 8, the average up to 10 is $(1 + 4 + 4 + 0 + 4 + 8 + 0 + 0 + 4 + 4 + 8)/10 = 3.8$." He pauses. "If you can solve this problem in two weeks, I will give you one of my detachable teeth as a souvenir. I'm told that trochophore teeth are worth thousands of dollars on the black market."

Difficulty Level: ✳ ✳ ✳ ✳

95 2, 271, 2718281

I had a feeling once about Mathematics — that I saw it all. Depth beyond depth was revealed to me — the Byss and Abyss. I saw — as one might see the transit of Venus or even the Lord Mayor's Show — a quantity passing through infinity and changing its sign from plus to minus. I saw exactly why it happened and why the tergiversation was inevitable but it was after dinner and I let it go.

— Sir Winston Churchill, in H. Eves, *Return to Mathematical Circles*

"Dorothy, I have a quick puzzle for you."

"Dr. Oz, can't a girl get any sleep around here?"

"I'll let you get to sleep the moment you tell me the significance of the following sequence."

2

271

2718281

271828182845904523536028747135266249775724709369995957496696762772407663035354759457l

Dorothy turns to Dr. Oz. "My, that sequence grows faster than a Kansas tornado spins."

"That metaphor is not very poetic, but it will have to do. You have one hour to tell me the significance of the sequence."

Difficulty Level: 🐜🐜

96 Android Watch

I think that there is a moral to this story, namely that it is more important to have beauty in one's equations than to have them fit experiment. If Schrödinger had been more confident of his work, he could have published it some months earlier, and he could have published a more accurate equation. It seems that if one is working from the point of view of getting beauty in one's equations, and if one has really a sound insight, one is on a sure line of progress. If there is not complete agreement between the results of one's work and experiment, one should not allow oneself to be too discouraged, because the discrepancy may well be due to minor features that are not properly taken into account and that will get cleared up with further development of the theory.

— Paul Adrien Maurice Dirac, *Scientific American*

Dorothy and Dr. Oz are touring the McConnell Air Force Base and Boeing Military Airplane Company just outside Wichita, Kansas. To their right are a series of airstrips positioned in a seemingly haphazard manner.

"Dr. Oz, will planes actually land on those runways?"

"No, this is a puzzle I have set up for the people of Wichita. Here, take a look at this map." Dr. Oz projects a diagram of various runways (Figure 96.1).

B.C. MANSFIELD

96.1. At what intersections should the androids be stationed?

"What is the minimum number of androids needed to be stationed such that they can keep an eye on all the runways just by rotating their eyes around their cylindrical heads? Where are the androids stationed? If you like, you can place androids in the white lane around the airport, but these are not runways, and, thus, the androids don't have to keep an eye on all of the white rectangular lane."

Difficulty Level: ✳ ✳

97 More Knight Moves

"Mathematics is a field in which one's blunders tend to show very clearly and can be corrected or erased with a stroke of the pencil. It is a field which has often been compared with chess, but differs from the latter in that it is only one's best moments that count and not one's worst. A single inattention may lose a chess game, whereas a single successful approach to a problem, among many which have been relegated to the wastebasket, will make a mathematician's reputation.

— Norbert Wiener, *Ex-Prodigy: My Childhood and Youth*

Dorothy wanders through one of Dr. Oz's testing facilities and finally finds him in a small, dimly lit office. Dr. Oz is typing with two of his tentacles as a smoking cigar dangles out of one of his breathing orifices. On the walls of the room are charts of Canada, the United States, and Mexico. Pins of various colors pierce some of the major cities. In an adjacent room, Dorothy sees several trochophores playing chess.

Dr. Oz motions for Dorothy to follow him, and soon they are in an elegant living room with a strange chesslike playing board etched into the tile floor (Figure 97.1). The board is cross-shaped and contains the numbers 1 through 13.

97.1. More knight moves. (Illustration by Brian Mansfield.)

"Dorothy, consider the following puzzle that appears to have been designed by a schizophrenic chess player. The object is to use a chess knight piece to start on any circle on the bottom row (with 1's) and finally arrive at any circle at the top row (with 13's). As an example, you can start at 1, then jump to 2, then 4, and so forth."

Dorothy sees that several trochophores are studying the same problem in the nearby foyer. A few of the creatures argue. One trochophore suddenly pokes another trochophore's abdomen with a tentacle. The shouting continues.

"Dr. Oz, what are the numbers for?"

"The sum of the numbers you land on must be exactly equal to the prime number 89. Can you find a path from bottom to top that sums to 89? Can you find a solution that passes through the special 'zero' circle in the middle to solve the problem?"

Difficulty Level: 🕷 🕷

98 Pool Table Gambit

In science one tries to tell people, in such a way as to be understood by everyone, something that no one ever knew before. But in poetry, it's the exact opposite.

— Paul Adrien Maurice Dirac, in H. Eves, *Mathematical Circles Adieu*

Dorothy is in a combination pool hall and country bar filled with aliens from several star systems. Dr. Oz motions Dorothy to a pool table (Figure 98.1). "Dorothy, my friends have devised a terribly difficult problem for you."

"Just what I need when I thought I had time to relax, sip ginger ale, and listen to Shania Twain sing a few tunes."

"You can relax after you solve the puzzle. Start at the cue ball – the white ball with no number on it – and draw a line to the ball marked '1'. Your line need not be straight. From the '1' ball draw a line to the '2' ball, and so on. You are not allowed to cross any of the lines you draw, and you can't draw a line that touches the rectangular frame enclosing all the balls. Your mission is complete when you draw the final line connecting '14' to '15'."

Dorothy hears a helicopter, and then she looks around the bar. Where is Toto? A bar is no place for a pet. She listens for Toto's nails tapping on the wooden floor, but all she can hear is the chop of helicopter rotors. The cacophony envelopes Dorothy, and she has to take deep breaths to remain cool.

"What do I get if I solve the problem?"

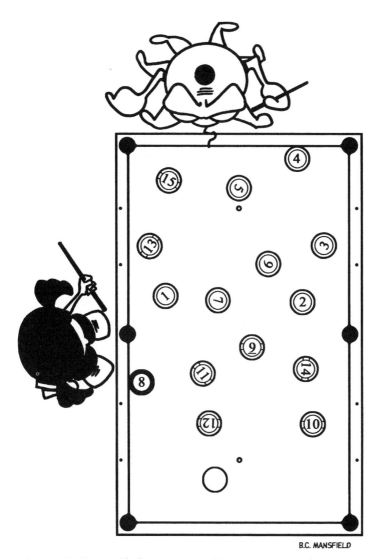

98.1. Pool table gambit. (Puzzle designed by Brian Mansfield.)

"A week's vacation on my home world. But you'll have to bring your scuba gear. Back home, we live mostly under water."

Difficulty Level: ✳ ✳ ✳

99 A Connection between π and e

God exists since mathematics is consistent, and the Devil exists since we cannot prove it.

— Andre Weil, in H. Eves, *Mathematical Circles Adieu*

Dr. Oz runs his tentacles over a map of the Earth. He looks pensive as his anterior tentacle rests on the globe and covers Florida.

"Dorothy, I'd like to consider connections between π (3.1415...), which is the ratio of a circle's circumference to its diameter, and Euler's constant *e* (= 2.7182...), which, like π, also appears in countless areas of mathematics. On our world, we feel that the most profound and enigmatic formula known to any life forms is the following." Oz draws an equation on a board:

$$1 + e^{i\pi} = 0$$

Dorothy looks at the equation. "Tell me more."

"Some believe that this compact formula is surely proof of a Creator. Others have actually called $1 + e^{i\pi} = 0$ 'God's formula.' Edward Kasner and James Newman in *Mathematics and the Imagination* wrote, 'We can only reproduce the equation and not stop to inquire into its implications. It appeals equally to the mystic, the scientist, the mathematician.'"

"Dr. Oz, who first discovered this formula?"

"Leonhard Euler, who lived in the 1700s. The formula unites the five most important symbols of mathematics: 1, 0, π, *e*, and *i* (the square root of –1). This union was regarded as a mystic union containing representatives from each branch of the mathematical tree: Arithmetic is represented by 0 and 1, algebra by the symbol i, geometry by π, and analysis by the transcendental e. Harvard mathematician Benjamin Peirce said about the formula, 'That is surely true, it is absolutely paradoxical; we cannot understand it, and we don't know what it means, but we have proved it, and therefore we know it must be the truth.' Notice the formula also contains the additive, multiplicative, and exponentiation operator, and that 0 and 1 are the two neutral units for these operators."

"Fascinating, but I sense this is leading up to one of your usual puzzles."

"Today I'm interested in a much more amusing question: What is the longest string of consecutive digits within π that have also been found in e? So far, the string 71828182 is the longest one I've found in both constants. Recall that e is 2.**71828182**84590452353602.... I found the string 71828182 in e also in π at position 58,990,555 counting from the first digit after the decimal point (the initial 3 in π is not counted). The string and surrounding digits in π are as follows: 17708342647565748477**71828182**93786843571860331854."

"What a delightful piece of trivia."

"Dorothy, here is your puzzle that relates to π and e. Of what relevance is this sequence?" Oz hands Dorothy a card with a number sequence:

6, 28, 241, 11,706, 28,024, 33,789, 1,526,800, ?

Mystery sequence

"Dorothy, also try to figure out what the next number is in the sequence. So far no Earthling has ever managed to solve this."

"Dr. Oz, that's difficult! Not even a hundred Albert Einsteins working with pencil and paper could solve this one. You had better give me an excellent reward if I solve this."

"Certainly, but you also have to solve another puzzle. It's the one I mentioned a minute ago: What is the longest string of consecutive digits within π that have also been found within in e?"

Difficulty Level: ✳ ✳ ✳ ✳

100 Venusian Number Bush

Mathematics as a science, commenced when first someone, probably a Greek, proved propositions about "any" things or about "some" things, without specifications of definite particular things.

— Alfred North Whitehead, *The Aims of Education*

Today Dorothy and Dr. Oz are exploring Venus. Dorothy looks at Dr. Oz through the helmet of his space suit, and she is once again struck by the glossiness of his huge eyes. Sometimes they seem to see into her very soul.

A tentacled alien approaches them. Over the speaker of Dorothy's spacesuit, she hears the alien's twangy voice.

"As you know, your Sun will eventually become a red giant and destroy Earth. However, our advanced technology can prevent this. You just have to prove that your species is worthy."

Dorothy steps back, unsure of where the alien is heading with his comments. She turns to Dr. Oz. "Is this fellow one of your colleagues."

Dr. Oz nods.

The other alien points to a crystalline bush and hands Dorothy a felt-tipped pen. "I have a test for you. Fill in numbers in such a way that the number in each circle is equal to the sum of the numbers in the circles into which it branches off. For example, the four circles into which the bottom trunk branches off should contain numbers that add up to a total of 102." The alien pauses. "I've already filled in three numbers. As you fill the circles, you must first use each of the numbers 1 through 12 once and only once, but after that you are free

100.1. Venusian number bush. (Illustration by Brian Mansfield.)

to fill the remaining circles with whatever numbers you like" (Figure 100.1).

Dorothy nods as Dr. Oz paces back and forth. He slowly gazes at jagged mountains perched besides an orange fjord illuminated by the evening sun.

The alien approaches Dorothy. "If you solve this within 24 hours, we will save Earth from eventual destruction."

Difficulty Level: ☀ ☀

101 Triangle Cave

Let us grant that the pursuit of mathematics is a divine madness of the human spirit, a refuge from the goading urgency of contingent happenings.

    – Alfred North Whitehead, in N. Rose, *Mathematical Maxims and Minims*

Back on Earth, Dorothy is cooking clam chowder on the stove and making sushi for Dr. Oz when she suddenly feels a great gust of wind coming from the nearby cave. Dr. Oz stands up from the table and shuffles to the cave entrance.

"Dorothy, this reminds me of some of my favorite puzzles." He hands her some cards. "Each of these puzzles represents a cave. Your job is to section off a part of the plot with a continuous rope. To lay the rope, draw a line through the triangle sides (not through the corners of a triangle), continuing until you return to where you start, forming a single closed loop. The digits in the triangles indicate the number of neighbor triangles that have a rope through them."

"What exactly is a neighbor triangle?"

"Neighbor triangles share sides. Look at the example." Dr. Oz hands Dorothy a card (Figure 101.1).

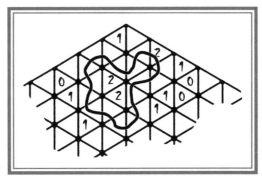

101.1. Example solved puzzle for the triangle cave.
(Illustration by Brian Mansfield.)

Dorothy looks into a subterranean lake and sees many tiny squid wriggling along the surface. She winces as the tentacled beings make a flatulating sound and then quickly dart away into the darkness.

"Dorothy, this example shows a solved puzzle. Notice that the ropes can go through either empty or numbered triangles. Now, I want you to focus on the unsolved puzzle." Dr. Oz hands Dorothy another card (Figure 101.2).

"Can you build the rope path in the unsolved puzzle? Does your path enclose the largest possible area of cave?"

101.2. Can you lay the rope in the triangle cave? (Illustration by Brian Mansfield.)

B.C. MANSFIELD

102 Rat Attack

By relieving the brain of all unnecessary work, a good notation sets it free to concentrate on more advanced problems, and, in effect, increases the mental power of the race.

— Alfred North Whitehead, in P. Davis and R. Hersh,
The Mathematical Experience

The trochophore party is filled with excitement. Many of the aliens jostle one another, talk, shout, and smoke a strange vermilion weed. Near one of the doors is a huge trochophore trying to twirl a Hula Hoop with his posterior tentacle.

There is smell on Dr. Oz's breath – something fruity and alcoholic. It reminded Dorothy of Aunt Em's apple pie.

"Dr. Oz, I'm of the opinion that one never should drink an alcoholic beverage."

"Dorothy, no need to worry. My biochemistry is such that I can never become drunk. In fact, alcohol interacts with my cerebellum in such a way as to permit me to create the most fascinating puzzles. Here is my latest."

Dr. Oz shows Dorothy a cage containing several rats. "The black rat (*Rattus rattus*) and the Norway rat (*R. norvegicus*) are ornery creatures that live with humans throughout most of the world and eat almost anything. The rats have a strong sense of smell and hearing, and their ability to burrow and gnaw lets them get through just about anything! Interestingly, when both species live in the same area, they occupy different habitats. In a building, for example, the Norway rat occupies the lower levels, while the black rat lives on the upper floors."

"Fascinating, Dr. Oz."

"And now to our puzzle. It is Friday the 13th, and you are alone in an old mansion created by an eccentric builder who made strangely shaped rooms." Dr. Oz hands Dorothy a diagram (Figure 102.1).

START

END

102.1. **Rat attack.** (Illustration by Brian Mansfield.)

"The illustration represents the mansion as viewed from above. Each line is a wall. The rats are so hungry that both the black and Norway rats are cooperating in their quest for food. The problem for the rats is to make as few holes as possible from the top of the figure to the bottom. (The rats want to make as quick an escape as they can after scarfing up their favorite foods in the mansion.) Where should the rats make their holes? No cuts can be made through the intersection points of walls, which are reinforced by steel to keep the walls from crumbling. Remember, the holes must be made from the top horizontal wall to the bottom horizontal wall."

Difficulty Level: 🕷

103 The Scarecrow Formula

Dorothy: "How can you talk if you haven't got a brain?"
Scarecrow: "I don't know. But some people without brains do an awful lot of talking, don't they?"

> — *The Wizard of Oz* (1939 MGM movie).

Dorothy is with Dr. Oz at a trochophore convention. She looks into the larger room permeated with the scent of pine accompanied by a strong medicinal odor. On the floor is a Persian rug depicting deer dancing through a waterfall. Big wooden boxes lean against the walls and are stacked to the ceiling. Strange words like "Bandages," "Enemas," "Latex Gloves," and "Forceps" are marked on the boxes in thick black letters.

Dr. Oz coughs to get Dorothy's attention. "Dorothy, in the 1939 movie *The Wizard of Oz,* there is a delightful Scarecrow who goes to see a powerful wizard to ask the wizard for some brains. After embarking on a long and dangerous journey, the great Oz gives the Scarecrow a diploma with an honorary degree of Th.D., that is, a Doctor of Thinkology."

"We didn't get movies on the farm, except those about agriculture. Plus the color on our old TV would come and go. *Wizard* sounds wonderful. And I find it curious that his name is the same as yours."

Dr. Oz nods. "Once the Scarecrow receives his brains, he immediately impresses his friends by reciting the following mathematical equation." Dr. Oz hands Dorothy a card:

> "The sum of the square roots of any
> two sides of an isosceles triangle is
> equal to the square root of the
> remaining side."

"Hold on," Dorothy says. "This sounds very similar to the Pythagorean theorem in mathematics, but it doesn't seem quite right. The supposedly brainy Scarecrow is not so smart after all."

"You are correct that something is wrong with the Scarecrow's formula, but can you tell me why his statement is wrong?"

Difficulty Level: 🦟

104 Circle Mathematics

Our minds are finite, and yet even in these circumstances of finitude we are surrounded by possibilities that are infinite, and the purpose of life is to grasp as much as we can out of that infinitude.

— Alfred North Whitehead, in N. Rose, *Mathematical Maxims and Minims*

Dr. Oz leads Dorothy into a great cathedral. "Dorothy, I'd like you to examine this design etched into the floor."

"It's beautiful!"

On the marble floor is a colorful pattern of circles within circles. Some of the circles are so tiny that Dorothy would need a microscope to visualize them (Figure 104.1).

"Yes, my assistants have repeatedly filled the spaces between tangent circles to create an elegant geometric pattern. Can you determine what the numbers in each circle signify?"

Dorothy is deep in thought but then notices tinkling noises coming from the priest's chamber. It sounds like the snapping of twigs or the breaking of small animal bones.

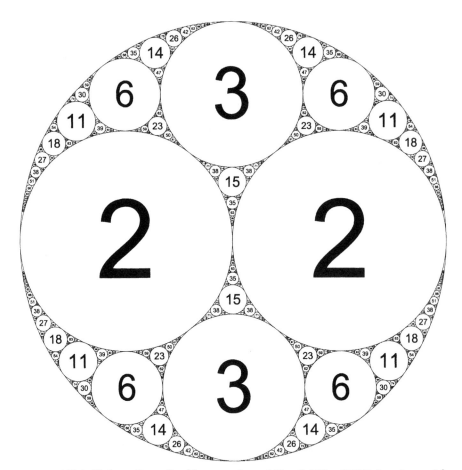

104.1. Circle mathematics. (Figure courtesy of Allan R. Wilks, AT&T Labs – Research.)

Difficulty Level: 🌋 🌋

105 *A, AB, ABA*

It is a profoundly erroneous truism, repeated by all copy books and by eminent people when they are making speeches, that we should cultivate the habit of thinking of what we are doing. The precise opposite is the case. Civilization advances by extending the number of important operations which we can perform without thinking about them.

       — Alfred North Whitehead, *An Introduction to Mathematics*

Dr. Oz leads Dorothy through tunnels that frequently and abruptly turn back on themselves, as if the Wichita facility had been periodically pummeled by artillery and quickly patched together again. The ceiling is a maze of wires, clamps, and dripping pipes. The walls are splotched with purple, red, and teal paint. Perhaps the painters had been in a hurry or did not care about their choices of color. Walking through the twisting tunnels puts Dorothy in mind of a journey through a vast, psychedelic intestine.

A huge alien with a bulbous head the size of a basketball approaches Dorothy and says, "Find the smallest number that, when divided by *AB, ABA,* and *ABAB* leaves the remainders *A, AB,* and *ABA,* respectively. (The letters *A* and *B* stand for digits of a number. For example, we might be talking about 8, 878, and 8787.) You have *AB* minutes to solve this problem or we will take over *AB* nations on Earth. As a hint, your solution may contain the consecutive digits 1976, the bicentennial of the United States. How do you solve this?"

"Sir, this problem is very, very difficult."

"No one from Kansas has ever solved it."

Difficulty Level: ✺ ✺ ✺ ✺

106 Ants and Cheese

To find the simple in the complex, the finite in the infinite – that is not a bad description of the aim and essence of mathematics.

> — Jacob Schwartz, *Discrete Thoughts*

Dr. Oz's five alien ant friends insist that their portions of cheese each have the same shape. Can you help teach them how to divide the cheese shown here (Figure 106.1) into five identical pieces?

106.1. Ants and cheese. (Illustration by April Pedersen.)

Difficulty Level: 🐜 🐜

107 The Omega Crystal

Picture a diamond a foot long. The diamond has a thousand facets, but the facets are covered with dirt and tar. It is the job of the soul to clean each facet until the surface is brilliant and can reflect a rainbow of colors.

— Brian Weiss, M.D., *Many Lives, Many Masters*

Dr. Oz points to an attractive assembly of attached boxes with everdiminishing sides. "Dorothy, this is my Omega Crystal" (Figure 107.1).

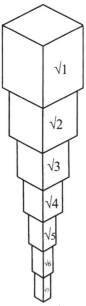

Etc.↓ $\sqrt{n}$ **107.1. The Omega Crystal.**

Dorothy steps closer to the Omega Crystal. "Remarkable!" she says as she examines the structure. The smallest boxes are so tiny that she would need a microscope to see them. "I wish I had a magnifying glass —"

"Never mind that. I want to tell you more about the boxes. The sides of the boxes diminish in an interesting series." Dr. Oz pulls out a slip of paper with the following series:

$$1 + \frac{1}{\sqrt{2}} + \frac{1}{\sqrt{3}} + \frac{1}{\sqrt{4}} + \ldots + \frac{1}{\sqrt{n}} + \ldots$$

Dr. Oz hands the paper to Dorothy. "The first box at the top of the crystal has an edge one foot in length. The next box has an edge length of one over the square root of two, and the next box has an edge length of one over the square root of three, and so forth. This series diverges, or gets bigger and bigger, which means that the Omega Crystal is a structure of infinite length! If you wanted to paint the faces of the Omega Crystal, you would need an infinite supply of paint."

A few visiting dignitaries – originally from the Virgo cluster of galaxies – approach Dr. Oz and Dorothy. The dignitaries' abdomens and thoraxes ooze a fluid that smells like roses on a warm spring day.

Dr. Oz bows to them and continues. "Remarkably, even though the length is infinite, the volume of the Omega Crystal is finite! What is the volume? If you can answer within two weeks I will give you this cherished crystal as a prize. If not, I will shatter it, which would then ignite a transgalactic war of unimaginable proportions."

One of the dignitaries moans and then evaporates.

Dorothy turns to Dr. Oz. "Really?" she says.

"Not really," Dr. Oz whispers. "I simply like to use bombastic language to impress our guests. Now get to work!"

Difficulty Level: 🕷 🕷 🕷

108 Attack of the Undulating Undecamorphs

If I am given a formula, and I am ignorant of its meaning, it cannot teach me anything; but if I already know it, what does the formula teach me?

— St. Augustine (354–430), *De Magistro*

Dorothy rocks back and forth as she stares at Dr. Oz's huge, spherical eyes. Crystals of happiness fill his eyes. His limbs are splayed. The large membrane encircling his nematocysts has ruptured. Perhaps he is in love.

"Dorothy, I have enjoyed our journeys around the solar system and to the farthest reaches of human imagination."

"Yes, they've been quite enlightening for me as well. Through your arduous training over the years, my mind has surpassed yours. My estimation is that I am now twice as smart as you. Now, I have a puzzle for you."

Dr. Oz steps back. "For me?"

Dorothy nods. "Dr. Oz, today I will talk about the exotic *undulating undecamorphic* numbers. As an appetizer, let's discuss polygonal numbers. The early Greek mathematicians noticed that if groups of dots were used to represent numbers, they could be arranged so as to form geometric figures." Dorothy draws a series of increasingly larger triangles:

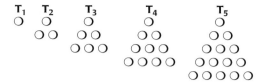

Because 1, 3, 6, 10, and 15 dots can be arranged in the forms of triangles, these numbers are called *triangular numbers.* The ancient Greeks were fascinated by these kinds of numbers."

"Dorothy, can we experiment with other polygonal numbers – hexagonal, for example?"

"You bet, Dr. Oz. Amazingly we can compute all polygonal numbers with a simple formula." Dorothy sketches on his tentacle:

$$\frac{r}{2}\left[(r-1)n - 2(r-2)\right]$$

"Here r is the number of sides of the polygon, and n is the 'rank.' The rank is simply an index that goes as $n = 1, 2, 3, \ldots$." Dorothy shows Dr. Oz a diagram showing the first three triangular, square, pentagonal, and hexagonal numbers (Figure 108.1).

Number	Rank		
	1	2	3
Triangle	●		
Square	●		
Pentagon	●		
Hexagon	●		

108.1. **Polygonal numbers.**

Dorothy reaches into a hidden compartment located within a chipped statue of a trochophore. She removes a notebook computer and tosses it to Dr. Oz. He types a short computer program that makes use of the formula. The program prints a few polygonal numbers:

Rank	1	2	3	4	5	6	7
Triangular	1	3	6	10	15	21	28
Square	1	4	9	16	25	36	49
Pentagonal	1	5	12	22	35	51	70
Hexagonal	1	6	15	28	45	66	91
Heptagonal	1	7	18	34	55	81	112

"Dr. Oz for the next half hour, we are interested in polymorphic numbers, and, in particular, undulating polymorphic numbers. Professor Charles Trigg (1898–1989) defined a number to be *polymorphic* if it terminates with its associated polygonal number. For example, hexagonal numbers have the form $H(r) = r(2r - 1)$. The number $H(125)$ is hexamorphic because $H(\mathbf{125}) = 31,\mathbf{125}$. Let's now define another mathematical term. The term *smoothly undulating integer* was used by Trigg in a paper on palindromic octagonal numbers, where he defines an integer as smoothly undulating if two digits oscillate – for example, 79,797,979. *Smooth* here differentiates this kind of number from an undulating integer where the alternating (i.e., adjacent) digits are consistently greater or less than the digits adjacent to them – for example, 4,253,612. For now, we are interested in undulation, not smooth undulation, because smoothly undulating polymorphic numbers are so rare that even computer searches become difficult."

Dorothy pauses as she reflects upon the trochophore colonization of Earth. They seem part of everyday life these days. The trochophores have established a flourishing trading relationship with humanity. The trochophores provide technologically advanced materials like nanotechnology, robotics, body probes, and Turing computers that have revolutionized Earth's industries. The aliens have also flown humans on tours of the Solar System, including the moons of Saturn. In return, humans told the trochophores about Earthly religions. For some reason, the trochophores are fascinated by the subject.

Dorothy continues. "An undulating polymorphic number can be defined in terms of its digits d." She draws on a board:

$$[(d_\lambda - d_{\lambda-1})(d_{\lambda+1} - d_\lambda)] < 0, \qquad \lambda = 2, 3, \ldots, N-1$$

"Here N is the number of digits in the number. Note that I define undulating polymorphic numbers as having an undulation in both $p(r)$ and r. This means that both the rank number and the associated polygonal number must oscillate. As an example, the following table includes the only undulating undecamorphic integer discovered for $r < 100,000$. (An 11-sided polygon is called an undecagon.)"

r	$p(r)$
25	2,7**25**
625	1,755,**625**
	(r undulates, but $p(r)$ "nearly undulates" due to consecutive 5's)
9376	395,55**9,376**
	(r undulates, but $p(r)$ "nearly undulates" due to consecutive 5's)

Undulating undecamorphic integers < 100,000

"Dr. Oz, can you find any other undulating undecamorphic integers, or is (**25**, 2,7**25**) the only pair representing such a beast? You have one day to answer me or I will cleave a one-inch portion from your anterior tentacle."

Difficulty Level: ✳ ✳ ✳ ✳

Epilogue

I love mathematics not only for its technical applications, but principally because it is beautiful; because man has breathed his spirit of play into it, and because it has given him his greatest game – the encompassing of the infinite.

— Rozsa Peter, *Playing with Infinity*

Many years have passed since Dorothy's initial escapades with Dr. Oz and Mr. Plex. As a result of her mental exercises and Dr. Oz's technology, Dorothy's intellectual capacity has grown exponentially. On the other hand, except for a few alterations, today Dorothy looks much as she did on Aunt Em and Uncle Henry's farm. She became immortal just a month ago by redesigning the biochemistry of her body and incorporated cybernetic implants and polymer prostheses.

Dorothy spends most of her time piloting a sleek spacecraft that Dr. Oz gave her as a gift. Now she is captain of her ship, always in command and sure of herself. She explores the universe along with Mr. Plex, Toto (whose surprising longevity she'd ultimately come to understand), and clones of Aunt Em and Uncle Henry, whose originals died long ago. Aunt Em is primarily an android, which gives her strength and durability for some of the more dangerous missions. Dr. Oz accompanies Dorothy and her motley crew for most of her missions exploring the cosmos.

Today, Dorothy guides her craft between enormous dust and gas clouds permeating the beautiful Orion Nebula, 1500 light-years from

231

Earth. Radiation pressure from the nebula's stars pushes nearby gas away, creating cavities within the clouds.

Dorothy gazes at her viewscreen. Four hot young stars, called the Trapezium, have hollowed out the nebula's core and touch the periphery in a few places. Dorothy sees new regions of uncharted space through the clear breaks scattered throughout the turbulent gases.

Dorothy turns to Doctor Oz. "I must apologize for Aunt Em," she says. "Her vocal subprocessors were not working correctly when she called you an 'odius exudate of infrahuman scunge.'"

Dr. Oz waves his tentacle. "Dorothy, think nothing of it. Aunt Em seems to be working just fine now."

Dorothy and Dr. Oz ascend a flight of stairs into an elegant living room. The rugs on the floor are oriental. A single massive table is draped in white linens. On one wall is an enigmatic sequence of numbers, the meaning of which no Earthling, except Dorothy, has ever deciphered:

3, 4, 20, 21, 119, 120, 696, 697, 4059, 4060, 23,660, 23,661, 137,903, 137,904 . . .

On another wall is a mysterious quadrilateral of which only Dorothy knows the significance (Figure E.1).

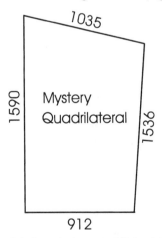

E.1. A mysterious quadrilateral. Can you determined the significance of this shape?

"Come here," Dorothy says to Dr. Oz as she motions him to sit near a violet-silk shaded lamp and a vase of roses near the center of the table. The silver, crystal, and china glitter like amethysts in the lamp's glow. On the far wall is a large map of Kansas and a photo of Dorothy's old farm.

Dr. Oz smiles. "I know where I want to go on our honeymoon," he says. "Take me to the Tarantula Nebula. We were there long ago. Or take me to the rings of Saturn or the moons of Mars."

There is a gasp from beneath the table. Aunt Em emerges from an estivation chamber beneath the table. "Honeymoon?" she says.

"Honeymoon?" Dorothy repeats with a grin.

Dr. Oz takes a ring from his pocket. Aunt Em raises an eyebrow and then opens up her mouth revealing row upon row of sparkling, quartz teeth. "Well," Aunt Em says, "I suppose you would make an acceptable husband, so long as you let me accompany you while you and Dorothy explore the stars."

Dr. Oz nods. "Dorothy, I know I was once evil, but I have since spent years proving that I have changed my ways. You have redesigned my brain and my genetic structure. I underwent this modification with pleasure, knowing that it might draw us closer and rid me of my malicious impulses. I am happier, and I am surely not the same being who treated you so harshly. Yes, I am different. This is literally true: I have even engineered my body to become more humanoid, and you have grown a tentacle to become more trochophorean." He pauses. "You have shown me that kindness and compassion are always a good course by which to steer. We are made for each other."

"Yes, that's true," Dorothy says. "We both enjoy mathematical puzzles and interstellar exploration."

Aunt Em walks over to a refrigerator and interrupts Dorothy. "What do you want to drink? I've got ginger ale."

"Shh!" Uncle Henry says as he enters the room. "Can't you see they need their privacy?"

They all return to the ship's bridge and command center, and Dorothy sit down in her captain's chair. Aunt Em takes the chair next to her.

Dr. Oz ambles closer toward Dorothy and looks into her eyes. "Will you be my wife?" he says as he slips the rhenium ring onto Dorothy's newly grown tentacle.

"Yes," Dorothy says.

"Hurray," Aunt Em says as she claps her powerful handlike appendages.

Dorothy smiles, shifts into neutral, and pulls Dr. Oz closer. Their tentacles coil about one another for a long, long time.

When Dorothy throws the craft into hyperdrive, she tells Dr. Oz, "Tonight, we'll find a planet to settle on, and when we do, you and I are going to be together on a wonderfully strange beach in a new world."

Dr. Oz grins.

Uncle Henry sighs.

Mr. Plex emerges from a portal in the floor and attempts to give a thumb's-up gesture.

Dorothy turns to Mr. Plex, Uncle Henry, and Aunt Em. "This is getting out of hand. I think I will have to make a new rule. No eavesdropping on honeymooners' conversations."

For a while, Dorothy, Dr. Oz, Mr. Plex, and her replicated family are quiet. Banking right, Dorothy coasts toward a planet. The planet grows darker, azure, more mysterious, as if it stands motionless in a watery cradle of time. But the planet is permeated with life. Bioluminescent aliens and robotic probes are so plentiful that the entire atmosphere is alive with colored twinkling lights. Soon it is as if Dorothy is looking at bright stars – like dazzling supernovas in a night sky. Sparkles are everywhere. Even in the planet's oceans. The whole world is a play of colors and bright arcs of light.

Dorothy initiates the landing sequence allowing them to descend toward this brave new world. She looks over at Toto and smiles. Toto barks.

Further Exploring

1. The Yellow-Brick Road

One approach to solving this problem is first to consider bricks of dimensions $8 \times 2 \times 4$ inches – a common brick size, according to the Bricklayer's Union of America. Next, consider highway lanes 12 feet wide. Fred Mannering and Walter Kilareski, in their blockbuster 1998 book *Principles of Highway Engineering and Traffic Analysis,* 2nd edition, suggest that 12 feet is the ideal width; however, they note that the width varies according to the terrain.

Although perhaps a bit wide for Dorothy's yellow-brick road, let's consider a four-lane highway and a distance of 2800 miles from New York City to Los Angeles. This long road will have a *length* of 2800 miles $\times$ 5280 feet/mile $\times$ 12 inches/foot = 177,408,000 inches (1.774080×10^8 inches). The *width* is 12 feet $\times$ 12 inches/foot $\times$ 4 lanes = 576 inches. Multiplying length times width we find that the road's area is 1.02×10^{11} square inches. (In this computation, we have assumed a flat and straight road, and it would be interesting to see how our final answer changes if one considers roads that must go up and down over mountains.)

Next we can compute how many bricks are needed to cover the road area of 1.02×10^{11} square inches. Note that the brick surface area is 8×2 inches if we assume the 8-by-2 side is face up. Therefore, the number of bricks required to pave the road is 1.02×10^{11} square inches $\div$ 16 square inches = about 6.39×10^9 or 6,390,000,000 bricks – roughly one brick for every human on Earth today.

We may compare this quantity of bricks to that which would be needed to construct the Great Pyramid of Egypt, the base of which is a square with 756-inch sides. The pyramid was originally 481 feet tall. We can compute the pyramid's volume thus: volume = area $\times$ height $\div$ 3 = 756 feet $\times$ 756 feet $\times$ 481 feet $\div$ 3 = 9.16×10^7 cubic feet or 1.58×10^{11} cubic inches. The number of bricks required to fill the Great Pyramid can be estimated by dividing the volume of the pyramid by the volume of an individual brick in the yellow-brick road, or 1.58×10^{11} cubic inches $\div$ 64 cubic inches $= 2.47 \times 10^9$ bricks.

We can compare the number of bricks in the yellow-brick road (6.39×10^9) to the number of bricks that would fill the Great Pyramid (2.47×10^9). The ratio 6.39×10^9 / 2.47×10^9 is approximately 2.5. This means that the bricks used to build the bicoastal yellow-brick road could also be used to construct 2.5 solid replicas of the Great Pyramid. In performing this calculation, Dr. Oz assumed no mortar between the bricks because the builders of the Great Pyramid did not use mortar. Also, L. Frank Baum does not indicate whether mortar was used in constructing the yellow-brick road in *The Wonderful Wizard of Oz*.

Think about the ratio 2.5. This ratio is notable because most people assume that paving a four-lane highway across the United States would require much more than two or three pyramids full of bricks!

Dr. Oz tried this puzzle on several colleagues who noticed that the term "America" was not clearly defined in the original puzzle. For example, one colleague assumed that South America was the continent of choice. Because South America has such a variable width, a 1000-mile long road was considered. This alternative computation is as follows. Assume a road about 11 feet 6 inches wide, about the width of Dorothy's road in the *Wizard of Oz* movie. The number of square feet needed to cover the road is therefore 1000 miles × 5280 feet/mile × 11.5 square feet of road to cover. We will consider a road one brick deep. For this alternative computation, let's assume the standard brick's longest two dimensions are 8.625 inches and 4.125 inches. In this calculation, we also assume that this surface is the one that tiles the road. This means that the road will be exactly 16 bricks wide. (If you want the traditional staggered effect, you need to chop a brick in half every other line, but it doesn't affect the count.) So we need 16 bricks × 1000 miles × 5,280 feet/mile × 12 inches/foot $\div$ 4.125 = 245,760,000 bricks. To be on the safe side, in case we need to build viaducts and bridges, this alternative computation yields an estimate of 300 million bricks, far less than the number of bricks in a single Great Pyramid. (Thanks go to my Internet teams, who worked collaboratively on this problem but used different numbers of significant digits for their calculations and slightly different brick sizes.)

With either estimate for the number of bricks, how tall a structure could you build that would be stable and last for a thousand years? While on the topic of stable brick formations, consider the following related problem. One day while walking through the Oz testing facility, Dorothy notices a stack of bricks leaning over the edge of a table. It seems as if it is about to fall.

Leaning tower of yellow bricks

A question comes immediately to mind: Would it be possible to stagger a stack of many bricks so that the top brick would be far out into the room – say 20 or 30 feet? Or would such a stack fall under its own weight? Dorothy asks several aliens, and each gives a different answer.

Simple as this question sounds, there have nonetheless been several discussions on this topic in prestigious physics journals. In this section I report on some of the remarkable findings, and encourage your involvement by including computer pseudocode. Fortunately for computer programmers, the simulation of the brick-stacking is relatively easy on a personal computer.

As Jearl Walker points out in *The Flying Circus of Physics,* the stack of bricks will not fall if the following rule is met: The center of mass of all the bricks above any particular brick must lie on a vertical axis that cuts through that particular brick. This must be true for each brick in the stack. Whether or not a stack falls can be determined by a computer simulation of a stack of bricks, where the centers of mass of the bricks are computed. This is easy to calculate. Assuming we are concerned only with a stack leaning in a single direction, the center of mass of a single brick is simply $(r + l)/2$, where r and l are the right and left coordinates corresponding to the edges of each brick. (For simplicity, we assume that each brick has the same thickness.) The composite center of mass for an entire stack of bricks is the average of all the individual centers of mass. The following computer recipe outlines the necessary computation. Users enter the right and left coordinates of each brick they wish to stack.

Test for the stability of a leaning stack of bricks

```
/* ------------- Data Entry ------------------------*/
Print("How many bricks do you wish to stack?");
Get(NumBrick);
Do i = 1 to NumBrick;
   Display("Enter left, right coordinates for brick number",i);
   Get (left(i),right(i));
end;
/* -------Will the stack topple ? ------------------*/
Do i = 1 to NumBrick-1
   /* initialize ctr. mass for all bricks above test brick */
   CmAbove=0
   Do j = i+1 to NumBrick
     CmAbove=CmAbove + (right(j)+ left(j))/2
   End /* j */
   /* compute composite center of mass */
   CmAbove=CmAbove/(NumBrick-i)
   If CmAbove > right(i) then do
     Display ("Stack topples")
     Print(i,CmAbove, right(i))
   end /* if */
end /* i */
```

To make the problem more interesting, the user can get an idea about how sturdy the stack is by providing small lateral vibrations to each brick and seeing if the stack falls. To do this simply add a small random number to each brick's center of mass. Execute the pseudocode several hundred times to see if the stack is still standing. For a slightly more complicated model, the displacement of each brick's movement can be simulated by a spring undergoing simple harmonic motion. Given a slight random force F applied to each brick, and a force constant k, the displacement of the brick's center of mass is given by $k = -F/x$. The value k need not be constant for each brick.

Many questions form in Dorothy's mind. Is it possible to make the stack jut out several feet? A mile? Is there any limitation as to how far the top brick can be beyond the edge of the table? Several papers in the *American Journal of Physics* and other journals point out that there is no such limit. For example, the top brick in a stack can be made to clear the table if there are five bricks in the stack (assuming identical brick sizes). For an overhang of 3 brick lengths, you need 227 bricks; for 10 brick lengths, you need 272,400,600 bricks; and for 50 brick lengths, you need more than 1.5×10^{44} bricks! Therefore, although there is no limit as to how far out one can "travel" with the brick stack, a great many bricks are required to do so. (We exclude complications such as Earth's gravity not being constant, the effect of the Moon, etc.) A formula for the amount of overhang attainable with n bricks, in brick lengths, can be used: $\frac{1}{2} \times (1 + \frac{1}{2} + \frac{1}{3} + \ldots + \frac{1}{n})$. This harmonic series diverges very slowly, so a modest increase in brick overhang requires many more bricks. This equation can be implemented in the following pseudocode:

ALGORITHM – Compute the harmonic series.
Determines the amount of overhang attainable with n bricks.

```
sum=0
Do i = 1 to n   /* n = number of bricks */
   sum = sum + 1/float(i)
end
sum = sum*0.5; Print(sum)
```

The first set of program code can be used as the foundation for several fascinating and educational computer games for students. For example, with a simple graphics interface two players can be presented a small group of bricks to stack. The player's goal is to make their stack (which appears on the screen) jut out as far as possible beyond the edge of the table without having the stack fall. They are each given a minute to arrange the bricks! For the scientifically minded experimenter, the program can display the exact point at which the stack was made to fall by graphically highlighting the "bad" brick and by displaying the various centers of mass. If you wish to take the game further, more complicated shapes can be used, such as triangles and circles. In another experiment, the computer can "throw" down bricks with random positions, and only those stacks that can physically stand up (as tested by the first program code) will be displayed. Throw several thousand bricks, and see what remains on the table.

Lots of papers have been written on the subject of leaning towers of objects. For example, see:

- Boas, R. (1973). "Cantilevered books," *American Journal of Physics* 41: 715.
- Johnson, P. (1955). "Leaning tower of Lire," *American Journal of Physics* 23: 240.
- Pickover, C. (1990). "Some experiments with a leaning tower of books," *Computer Language* 7(5) (May): 159–60. Also see Pickover, C. (1991). *Computers and the Imagination*. New York: St. Martin's Press.
- Sutton, R. (1955). "A problem of balancing," *American Journal of Physics* 23: 547.
- Walker, J. (1977). *The Flying Circus of Physics*. New York: Wiley.

2. Animal Array

The answer is 🐢 🐢. If a cell has just one side thickened, the cell gets a ➤ symbol. If the cell has two sides thickened, it gets a 🦋 symbol. If it has three sides thickened, it gets a 🐢 symbol. If it has four sides thickened, it gets a 🐢 symbol. Puzzles like these are a stimulus for the mind because there are so many different ways to attack such problems. Can you justify a different answer using different criteria?

Here is another challenging puzzle for you to ponder with friends. Fill in the empty tile with the correct missing symbols.

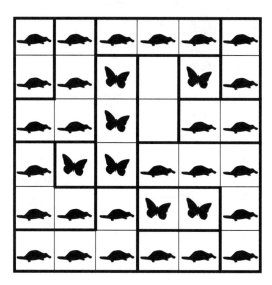

3. An Experiment with Kansas

Obviously, the smaller the region of the land on the globe, the lower the chances of Dorothy being able to hit the land with a random throw. For example, it's a lot easier to hit Asia than Kansas. We can very roughly approximate the solution by comparing the area of Kansas – 213,109 square kilometers (82,282 sq. mi.) – to the area of Earth (197 million sq. mi.). Thus the ratio of the area of Kansas to that of Earth is 82,282/197,000,000 = 0.000417 or 0.04%. Because Earth's area is about 2394 times the area of Kansas, Dorothy has roughly a 1 in 2,394 chance of hitting Kansas on her first try – if we model the problem as one in which Dorothy randomly throws a dart at the globe. Not very good odds! Her odds are slightly increased because she is not throwing a dart but rather an object shaped like Kansas. Unlike a dart, the midpoint of her tossed Kansas need not even fall on Kansas. You might approximate the answer by modeling Kansas as circle of radius r. This means that if the tossed Kansas's center falls within r miles of the edge of the Kansas on the globe, this counts as a hit. Can you determine by what amount this more sophisticated model increases Dorothy's odds of hitting Kansas?

Note that the surface area of habitable land on the globe is significantly less than Earth's overall area. There are about 58 million square miles of land, of which 49 million are habitable.

The experiment with Kansas becomes more complicated if we wish to consider it in detail. For example, the thrown object is more likely to hit the equator than near the poles – assuming that the globe spins on a vertical axis and we are launching Kansas from the side of the globe. (Our first analysis is actually more appropriate for a globe randomly bouncing around in a box that contains a sticky Kansas waiting to adhere to its curved surface.) However, if we assume that all parts of the globe have equal probability of being hit, and we ignore Earth's curvature, we could, in principle, use Blaschke's theorem in geometric probability that implies that the probability of the thrown Kansas touching the globe Kansas depends only on the area and perimeter of Kansas, and not on other aspects of its shape. This analysis is beyond the scope of *The Mathematics of Oz* but would surely impress Dr. Oz should you decide to pursue it further! (Wilhelm Blaschke [1885–1962] was a German mathematician who researched numerous areas in the field of geometry.)

4. An Experiment with Signs

Either the third sign or the last sign in the fourth row does not belong, because this sign is the only sign that is repeated in the collection.

5. The Logic of Greenness

One of the easiest ways to start attacking this problem is to run several experiments to convince yourself of the answer. For example, consider the following 4×5 rectangle of randomly placed greenness values ranging from 1 to 20. The higher the number, the more green the alien.

1	11	3	17
4	2	16	18
20	7	15	12
8	19	6	5 (C_{gp})
10	9	14 (R_{pg})	13

👽 **Aliens of varying greenness** 👽

First, we can consider the greenest of the palest aliens in each column. We'll denote this number by C_{gp}, which means "greenest of the palest in each column." To arrive at C_{gp}, let's list the palest in each column: 1, 2, 3, and 5. Therefore the greenest of these is $C_{gp} = 5$. Now let's calculate the "palest of the greenest aliens in each row," R_{pg}. First, let's find the greenest in each row: 17, 18, 20, 19, 14. Therefore, the palest of the greenest of these is $R_{pg} = 14$. We can see from this one experiment that $R_{pg} \geq C_{gp}$. It turns out that "value" of the palest of the greenest aliens in each row is always greater than or equal to the greenest of the palest in each column. Try the experiment yourself for many arrays of aliens. Impress your friends with the fact you always know the outcome.

Note that we can approach this from a more general perspective. Recall that the palest of the green is R_{pg}, and the greenest of the pale C_{gp}. Now look at the alien where the R_{pg}'s row crosses C_{gp}'s column. Clearly this alien at the intersection is at least as green as C_{gp} but no greener than R_{pg}. This implies more generally that $R_{pg} \geq C_{gp}$.

Can you extend your logic from a rectangular array to other array shapes or to higher dimensional shapes? What do you find?

My colleague Dharmashankar Subramanian provides a more analytical solution for those of you with a mathematical background. We may consider the greenness question as a question about the statement of weak duality of nonlinear programming. For the rectangular array, Subramanian asks us to index the rows and columns as follows. Let us index the rows as x(row$_k$ is x_k) belonging to set X. Let us index the columns, as y(column$_i$ is y_i) belonging to set Y. Let $F(x,y)$ be the function denoting "greenness." In the rectangular array, every (i, k) pair (where x_k belongs to X and where y_i belongs to Y) will have a value given by $F(x_k, y_i)$ that denotes the "greenness" of the alien-entry at the (i, k)th position in the array. Our greenness question might be asked more rigorously, for example: Is max(over Y) [min(over X) $F(x,y)$] $\leq$ min(over X) [max(over Y) $F(x,y)$]? The implication of weak duality suggests that max-over-Y (min-over-X $(F(x,y))$) $\leq$ min-over-X

[max-over-$Y(F(x,y))$]. We can prove this as follows. We know that, for any general x' belonging to X and y' belonging to Y:

(a) min-over-X $(F(x,y)) \leq F(x',y)$ [note that x would be that x that minimizes $F(x, y)$, y fixed at y']

(b) $F(x',y) \leq$ max-over-$Y[F(x',y)]$ [similarly, note that y would be that y that maximizes $F(x',y)$, x fixed at x'].

Therefore for any (x',y) we have min-over-$X[F(x,y)] \leq$ max-over-$Y[F(x',y)]$. This holds for any x' and y'. So, in particular it holds true for that y' that maximizes min-over-$X[(F(x,y)]$. Therefore, max-over-$Y[$min-over-$X(F(x,y)] \leq$ max-over-$Y[F(x',y)]$. This holds true for any x'. So, in particular, this holds true for that x', that minimizes max-over-$Y[F(x',y)]$. Therefore, max-over-$Y[$min-over-$X(F(x,y)] \leq$ min-over-$X[$max-over-$Y(F(x,y)]$. The palest in each column, say column$_i$, is min-over-$X[F(x,y_i)]$. The greenest of (palest in each column), over all columns, is max-over-$Y[$min-over-$X[F(x,y)]]$. Similarly, the greenest in each row, say row$_k$, is max-over-$Y[F(x_k,y)]$. The palest of (greenest in each row), over all rows, is min-over-$X[$max-over-$Y(F(x,y)]$. Therefore, the greenest of (palest in each column), over all columns, is less than or equal to the palest of (greenest in each row) over all rows. The result is true for continuous sets X and Y. In the rectangular array, X and Y are discrete sets.

Note that the green-alien problem has connections to game theory, a branch of applied mathematics that deals with scenarios in which there is a competition between players who have their own goals and often try to win by anticipating the opponent's decisions. For example, pretend we are playing a game using the aliens. I chose a row, and you choose a column, and we select the alien in my row, your column. You're trying to get the greenest alien you can get, and I'm trying to get the palest alien. If I go first, you'll pick the greenest alien in my row. I know you're going to do that, so I pick the row with the palest greenest alien. We wind up with R_{pg}. If you go first, we'll wind up with C_{gp}. Going second must be an advantage, right?

6. Magical Maze

Here is one solution. Can you find others? What is the most efficient way of solving such a problem?

(O)   Start.

🐇   Add 5 pebbles to your pocket.

🐾   Jump to any bird.

🐕   Add 30 pebbles to your pocket and jump to a dog.

🐐   Do not subtract pebbles whenever you are told to do so.

🦋   Subtract 5 pebbles and go to the start. (After Start and "Add 5 pebbles to your pocket," jump to the final bird.)

🐿   If you have over 38 pebbles, go to the next squirrel.

🦋   If you have an odd number of pebbles, go to a butterfly.

(O)   End here! Congratulations. You did it!

7. Kansas Railway Contraction

We can use the following formula to estimate the change in length for the track due to thermal expansion: $\Delta L = aL\,\Delta T$, where a is a coefficient of thermal expansion for gray cast iron, and ΔL and ΔT are the changes in length and temperature. This gives roughly $\Delta L = 6.8 \times 10^{-6} \times 2800$ miles $\times$ 130 degrees F or approximately 2.5 miles. Most people don't imagine that the track contracts or expands by a walloping 2.5 miles.

As an additional challenge, if we could somehow fix the two ends of the 2800-mile track so that when it expands it buckles into an arc, how high above the ground will the track be at its midpoint?

As for the chapter's second question, if you apply heat to the iron doughnut, it will expand but keep its same proportions. Therefore, the hole will get larger. This same logic applies to heating metal caps on jars. Of course, from a practical standpoint, the doughnut might not expand sufficiently to help Dorothy escape, and she would also have the challenge of not burning herself in the process. Don't try this experiment at home!

8. The Problem of the Bones

Of all the problems in this book, this one generates the most controversy. It is actually quite difficult, and I have found that many colleagues disagree on the best answer to give the alien if forced to provide a "best guess" for the ratio of the long piece to the short piece. What follows is a potpourri of different approaches, many of which have been stimulated by discussions with colleagues.

One way to attack this problem is to consider mathematical theory for finding an "average value" for the ratio of long piece to short piece. Pretend we have a "unit" bone that stretches between 0 and 1 on the number line. The breakpoint is at a random position x, which leaves us with one bone piece of length x and another piece of length $1 - x$.

Broken bone

To compute the length ratio of one piece to the other piece, we want to compute the average ratio of $1 - x$ to x. Unfortunately, it turns out that there is no average value for the ratio! It's a little tricky to understand precisely why. Let's give it a try. The average or mean of a set of values is found by adding up the values and dividing by the number of values. The mean value of a *function* is the generalization of this idea. You can visualize the mean value of any function $f(x)$ as the mean "height" of the function when it is graphed. Perhaps we wish to find the mean value of a function between two values of x, say between $x = a$ and $x = b$. This mean value can be found by calculating the area under the function and dividing by the "width" $b - a$. We shall denote the mean value of a function f as $\langle f \rangle$. For those of you who know a bit of calculus, one can traditionally attempt to find an average by performing the following integration:

$$\langle f \rangle = \frac{1}{(b - a)} \int_a^b f(x)\, dx$$

(We are assuming that the function is defined from a to b. In our case of the unit bone, $b = 1$ and $a = 0$.) More specifically, to find the average ratio of the two pieces of bones, we can compute the following:

$$\int_0^{0.5} \frac{1 - x}{x}\, dx + \int_{0.5}^1 \frac{x}{1 - x}\, dx$$

This equals

$$2\int_0^{0.5} \frac{1 - x}{x}\, dx$$

or $2[(\ln x - x)]$ from 0 to 0.5, but this quantity diverges as x approaches 0. Thus, there can be no average ratio of bone lengths!

✳   ✳   ✳

Here is a great opportunity for a short lesson in random variables. Obviously, the outcome of an experiment need not be a number; for example, the outcome of an insecticide application experiment might be either "dead insect" or "alive insect." However, it is often desirable to represent outcomes as numbers. A *random variable* is a function that associates a unique numerical value with every outcome of an experiment. Each time a researcher repeats an experiment, the value of the random variable may vary. To understand best how to calculate the *expected value* of a random variable, consider that the mean of a random variable (another way of saying its expected value) is the weighted average of all the values that a random variable assumes in the long run. An example will clarify.

We can compute the expected value of a random variable by making the variable assume values according to its probability distribution. Next we record all the values and calculate the mean. If we do this again and again, the calculated mean of the values will approach some finite quantity. If the mean does not approach a finite quantity, the mean is said to diverge to infinity, and the mean of the random variable does not exist.

Let's denote the expected value of a random variable Z as $A(Z)$. For a discrete random variable, $A(Z)$ is calculated from

$$A(Z) = \sum_z z p_z(z)$$

As an example, let's consider dice. Out of the 36 possible rolls of the dice, there is only one way to get a "2" (by throwing two 1's) and one way to get a 12 (by throwing two 6's). The random variable Z has the following probability distribution for $(z, p_z(z))$:

z	$p_z(z)$
2	1/36
3	2/36
4	3/36
5	4/36
6	5/36
7	6/36
8	5/36
9	4/36
10	3/36
11	2/36
12	1/36

The random variable Z assumes a value equal to the sum of two die rolls. Its expected value $A(Z)$ is calculated as

$$A(Z) = \sum_{Z=2}^{12} z p_z(z)$$

$= 2(1/36) + 3(2/36) + 4(3/36) + 5(4/36) + 6(5/36) + 7(6/36) + 8(5/36) + 9(4/36) + 10(3/36) + 11(2/36) + 12(1/36) = (1/36) (2 + 6 + 12 + 20 + 30 + 42 + 40 + 36 + 30 + 22 + 12) = (252/36) = 7$. Therefore, after many experiments, the average value of two die rolls using traditional dice is 7.

More information can be found at Thinkquest, Inc., "Expected value of a random variable," http://library.thinkquest.org/10030/5rvevoar.htm.

✳   ✳   ✳

Let's return to the problem of the bones. Just because the mean of the ratio of bone parts is not defined, all is not lost – because there are other useful definitions of mean, central tendency, and maximum likelihood that you may wish to explore. Therefore, we might be able to come up with a reasonable answer for the alien interrogator. Consider what you would do if *forced* to provide an answer to the alien. For which of the following definitions can we find a measure of central tendency that might satisfy the aggressive alien?

My preference is to use the "harmonic mean" when formulating an answer for the Bone Being. The harmonic mean is a perfectly good measure of central tendency and sure to win the alien's favor given the traditional arithmetic mean is not computable. What is the harmonic mean all about? Let's start with an example. For the given two quantities, 20 and 30, the number $(20 + 30)/2 = 25$ is known as their traditional *arithmetic mean;* the *harmonic mean* of the two numbers, however, is $2/(1/20 + 1/30) = 24$. More generally, we can define the harmonic mean as follows, if none of the numbers is 0:

$$A_h(a_1, a_2, \ldots, a_N) = N/(1/a_1 + 1/a_2 + \ldots + 1/a_N)$$

You can compare this with the definition of arithmetic mean:

$$A(a_1, a_2, \ldots, a_N) = (a_1 + a_2 + \ldots + a_N)/N$$

Another possibility is to use the *geometric mean:*

$$A_g(a_1, a_2, \ldots, a_N) = (a_1 \times a_2 \times \ldots \times a_N)^{1/N}$$

For our example of two quantities, 20 and 30, the geometric mean A_g is $(20 \times 30)^{1/2} = 24.49$. You can see for this example that all three means are quite similar: 25, 24, and 24.49.

For our bone problem, the *geometric* mean approaches 4 (this value can be obtained using computer simulations or integral calculus). Simulations suggest that the *harmonic* mean converges to $1/(2 \times \ln(2) - 1) = 2.588699\ldots$ as the number of bone breaks approaches infinity. If the alien is breaking $n = 10,000$ bones, my colleagues and I find the solution is 2.59 ± 0.04.

All of these different kinds of means are viable "averages," if we use the word generally to convey a "typical" set of values. Luckily, even though the arithmetic mean is unusable, both the harmonic and geometric means settle to precise values as the amount of data increases. So, 2.588 might be a good "ratio" to tell the bone alien if we wanted to demonstrate our mathematical prowess. One reason the harmonic mean seems to settle, or converge, is that we are taking reciprocals that produce values that naturally fall between 0 and 1. Consequently, as the number of values increases, the mean of their reciprocals usually converges to a precise finite limit. Taking the reciprocal of the result then transforms the values back to the original interval of 1 to infinity.

There are additional kinds of measures you may wish to explore. For example, the *root mean square* (RMS) value of a function is defined as the square root of the mean value of the square of a function. To compute the RMS value of a function we square the function, calculate the mean, and then take the square root:

$$f_{RMS} \equiv \sqrt{\langle f^2 \rangle} = \frac{1}{(b - a)} \int_a^b f^2(x)\, dx$$

Do you think this measure of central tendency diverges? Give it a try. You can probably think of many other measures of maximum likelihood that neither I nor Dr. Oz have ever considered.

David T. Blackston suggests that if R is our ratio of bone parts, and x is our breakpoint along the bone, then the probability of obtaining a ratio less than 3 ($R < 3$) is 50%. This is the probability associated with ($0.25 < x < 0.75$). This means Blackston would choose a ratio of 3 in the wager with the alien.

✳   ✳   ✳

Note that if you were an archaeologist examining the pit of bones after the fact, and not doing a mathematical analysis, you could not:

1) arbitrarily select any two bone fragments from the pit,
2) divide their lengths, and
3) repeat the process, hoping that the average ratio would be the same as if you had picked up the bones in some other order.

For example, consider five bones as follows:

Fragment 1	Fragment 2	Ratio
2	3	0.67
4	1	4
1	1	1
1	3	0.33

Ordering 1 gives mean ratio: 1.49

Fragment 1	Fragment 2	Ratio
2	1	2
4	3	1.33
1	1	1
1	3	0.33

Ordering 2 gives mean ratio: 1.16

This demonstrates that the mean value is not invariant with respect to how the bones are matched.

✳   ✳   ✳

A number of respondents performed simulations of the bone-breaking problem – not with human leg bones but with computers. Of course, our mathematical studies and simulations deal with idealized bones. For example, real bones are bigger near the ends and may be less likely to break near the ends when thrown against a rock. In fact, if you were to examine the bones in the pit and discover that there were few breaks near the ends, you might be able to model this feature mathematically and avoid some of the divergent mean calculations involved in some of the previously mentioned calculations.

If you wish to perform some bone-breaking simulations in the safety of your home, here is some BASIC code for computers.

```
5 REM N is the number of bones the alien shatters
10 N = 10000
20 T = 0
30 FOR C = 1 TO N
35 REM P is a random value between zero and one
40 P = RND
50 IF P >= 0.5 THEN L = P ELSE L = 1 – P
60 S = 1 – L
70 T = T + (L / S)
80 NEXT C
90 PRINT T / N
100 REM Thanks to Ed Murphy
```

Notice that the average ratio is increased by a breakpoint near the end of the bones more than it is decreased by a breakpoint equally near the middle.

Example break near the bone's end: $0.99/0.01 \rightarrow R = 99$
Example break near the bone's middle: $0.51/0.49 \rightarrow R = 1.04$

Does this mean that the more bones we examine by simulation, the higher the average ratio? As the number of bones approaches infinity, does the average ratio approach infinity? Perhaps this is the divergence that the initial integration formula for mean is measuring.

*   *   *

Darrell Plank writes to Dr. Oz:

I think many of these approaches are fraught with difficulties. One might conclude that there [is] a continuum of "averages," because one could choose to sum the zth powers of the numbers and then take the zth root. Unfortunately, by varying z, one could obtain many different results. The RMS case uses $z = 2$. The traditional mean uses $z = 1$. Yes, there are a number of means, but if you allow yourself to "pick and choose" after the fact, you can obtain different numbers for the mean. I think the proper way to approach this problem is to assume that the ratio can't be smaller than some very small number. For example, if you assume that the alien can't break the bone in a ratio less than 1/100 of the standard mean, this gives a nice 6.844 as an average. You ask, "Does this mean that the more bones we examine by simulation, the higher the average ratio?" Absolutely, assuming you're performing infinite precision simulations. However, with a typical computer's precision, you're going to get a much smaller number than infinity.

✳   ✳   ✳

Dr. Robert Stong has suggested that instead of trying to find the expected mean value of the ratio of the larger piece to the smaller piece, we might seek the expected value of the ratio of the smaller piece to the larger piece. Interestingly, this is a bounded function, so it should have an expected value. In fact, the expected value is

$$2\int_0^{0.5} \frac{x}{1-x} dx \quad \text{and}$$

$$2\int_0^{0.5} \frac{x-1+1}{1-x} dx = 2\int_0^{0.5} \left(-1 + \frac{1}{1-x}\right) dx$$
$$= 2\{-x - \ln(1-x)\}_0^{0.5} = 2\{-0.5 - \ln(1/2) + 0 + \ln 1\}$$
$$= -1 - 2\ln(1/2)$$
$$= 2\ln 2 - 1$$

Stong notes that 1 over this quantity is the harmonic mean of the ratio we sought, which is why simulations suggest that the harmonic mean converges to $1/(2\ln 2 - 1)$.

9. Square Overdrive

We can first attack the two problems involving square and cube numbers. *Problem 1:* What number added separately to 100, 200, and 300 will make the three results three different square numbers? The fiendish Dr. Oz has tricked Dorothy. No integer solution exists. We can show this by recasting the problem as a set of three equations:

1) $100 + x = a^2$
2) $200 + x = b^2$
3) $300 + x = c^2$

Here, x is the unknown number to be added to 100, 200, and 300 to produce the square numbers a^2, b^2, and c^2. Equation 2 minus equation 1 yields $100 = (b - a) \times (b + a)$. Stop and think for a moment. We know that 100 has only a limited number of factors, namely 1×100, 2×50, 4×50, 5×20, and 10×10. Through a little trial and error involving this limited selection of possibilities, you can show that there is no solution that will make a, b, and c integers. Additionally, I believe there are no solutions using rational numbers instead of integers. A *rational number* is a number that can be expressed as a ratio of two integers. Here are some fine examples: $\frac{1}{2}$, $\frac{4}{3}$, $\frac{7}{1}$, 8. All common fractions and all expansions with terminating (or repeating) digits are rational. In other words, $\frac{1}{2}$ = 0.5 is a rational number; likewise, so is $\frac{1}{7}$ = 0.142857142857142857142857, even though the digits repeat "1428571" ad infinitum. Trigonometric functions of certain angles are even rational, for example, cos 60° = $\frac{1}{2}$. (This is in contrast to irrational numbers like e and π – called *transcendental numbers* – and all surds such as $\sqrt{27}$. A *surd* is a number that is obtainable from rational numbers by a finite number of additions, multiplications, divisions, and root extractions.) The number π is represented by an endless and nonrepeating string of digits.

Problem 2: What number added separately to 100, 101, and 102, will make the three results three different cube numbers? This problem requires finding three cube numbers a^3, b^3, and c^3 so that:

1) $100 + x = a^3$
2) $101 + x = b^3$
3) $102 + x = c^3$

If you are clever enough to remember that numbers like -1, 0, and 1 are cube numbers, then one solution is $a = -1$, $b = 0$, $c = 1$, and $x = -101$.

As for *Problem 3:* Dr. Oz and Dorothy have little information on squarion arrays and welcome feedback from readers. Can they be made as large as desired? No one knows for sure. (Admittedly, the ✳✳✳ difficulty rating may be a bit too easy for Problem 3, as this problem may have no solution.)

While on the topic of square numbers, I cannot help mention "tridigital squares," researched by mathematician Joe K. Crump (http://www.space-fire.com/numbertheory/). *Tridigital squares* are square numbers that contain at most three unique digits. Here are all known tridigital squares with the digits 2, 6, and 9. Will humanity ever find a larger beast?

9
29929
69696
929296
9696996
996222969
26629996969
69926926969
269996262622969
9222222699262629962929
9929662926692269969296
9629629996292622299669929
2669292996662992629666262969
2699929699222292222296696996
62969966996922666266229669292996

On a related topic, I have a particular penchant for an unusual class of numbers called *exclusionary squares*. Can you tell me what is so special about the following number?

639,172

It turns out that this is the largest integer with distinct digits whose square is made up of digits not included in itself:

$639,172^2 = 408,540,845,584$

This particular exclusionary square was discovered by Andy Edwards of Queensland, Australia. Can you find the only other six-digit example? Can you find any *exclusionary cubes*? (I am unaware of any large exclusionary cubes, and we may wish to generalize this to exclusionary numbers of the *N*th order.)

10. Squares and Cubes

The fiendish Dr. Oz is asking us to find three integers, x, y, and z, such that:

1) $x^2 + y^2 + z^2 = N^3$

2) $x^3 + y^3 + z^3 = M^2$

One way to approach this problem is to simplify the two equations in order to reduce the number of variables. For example, let's try looking for solutions of the form $(-a, 0, a)$, all of which immediately satisfy equation 2, which reduces to $-a^3 + a^3 = M^2$, and $M = 0$. As we step through integer values of a, we quickly find a solution that satisfies equation 1, namely $-2^2 + 0^2 + 2^2 = 2^3$. Therefore, $x = -2$, $y = 0$, and $z = 2$ is one viable solution. Are there an infinite number of solutions? What is the best way to show the solution space using computer graphics?

On a somewhat related topic, equations of the form

$$a_1{}^k + a_2{}^k + \ldots + a_m{}^k = b_1{}^k + b_2{}^k + \ldots + b_n{}^k$$

with $a_1 \geq a_2 \geq \ldots \geq a_m;\ b_1 \geq b_2 \geq \ldots \geq b_n;\ a_1 > 1;\ m \leq n$

have fascinated me for a long time. Here, k, m, n, and terms a_i, b_j are positive integers. Don't you wonder, for example, if a solution exists for $a_1{}^6 + a_2{}^6 + a_3{}^6 = b_1{}^6 + b_2{}^6 + b_3{}^6$? Those of you wishing to make new discoveries in this area will be interested in the EulerNet project, a gigantic Internet search project that catalogs all the known minimal solutions to equations of these kinds. For example, Nuutti Kuosa, a member of the project, discovered this beauty in 2001:

$$1307^7 + 857^7 + 618^7 + 400^7 = 1184^7 + 1133^7 + 1030^7 + 423^7$$

The sum is equal to 6,890,807,721,574,272,667,868. For more information, see Jean-Charles Meyrignac, "Computing minimal equal sums of like powers," http://euler.free.fr/index.htm. Thousands of hours of computer computations are currently being exploited to make new discoveries.

On a related topic, consider that most positive integers can be expressed as the sum of three square numbers. For example, $9 + 9 + 9 = 27$ or $16 + 1 + 1 = 18$. Can you think of any numbers that can't be written as the sum of three or fewer squares? For example, what numbers require a sum of four squares to produce the numbers? In 1770, French mathematician Joseph-Louis Lagrange proved that every positive integer is either a square itself or the sum of two, three, or four squares. For further information, see Ivars Peterson, "Surprisingly square," *Science News* 159(24) (June 16, 2001): 382–3.

Recently, Len Stubbs from Victoria, Australia, challenged me to find any positive integer solutions to

$$A^n + B^2 = C^2$$

For example, two such solutions are $7^6 + 8400^2 = 8407^2$ and $6^7 + 23,325^2 = 23,331^2$. How many other solutions can you find? Are there solutions with smaller values for A, B, and C?

11. Plex's Matrix

Although the rules of replacement are simple, I have never found a person able to solve this puzzle without Dr. Oz's hint, and I doubt you will be able to find anyone who can supply a solution without the hint. One possible solution is to assign values to the symbols as follows:

$$\mathsf{F} = 1, \quad \mathsf{B} = 2, \quad \text{\reflectbox{?}} = 3, \quad \text{and} \quad \mathsf{M} = 4$$

In each row, the number assigned to the rightmost symbol is equal to the number assigned to the first symbol minus the second, plus the third, minus the fourth. Therefore, one solution for the missing symbol is F.

4	3	2	1	2
3	1	2	1	3
3	2	4	1	4
2	2	2	1	1
1	1	4	2	2

12. Chaos in the Clock Factory

Figure F12.1 shows what we believe to be the longest path. I do not know if it is possible to solve the puzzle if the clocks' hands are all rotated by 90, 180,

F12.1. Clock solution.

or 270 degrees – nor do I know if it is possible to solve the maze if each time you land on a clock it then rotates 90, 180, or 270. Is solving the puzzle simplified by allowing travel only in the directions of the hands on the clock being traversed?

Much more challenging games can be pondered if the clock factory puzzle is implemented as a computerized version. For example, every 15 seconds, the hands might rotate. Or, every time you land on a clock face, all adjacent clocks rotate. Alternatively, every three moves, the board grows in size, and

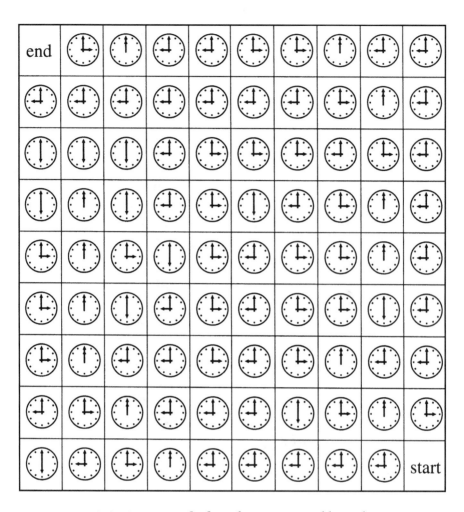

your mission is to get as far from the start as possible until you arrive at a dead end.

Here is a final clock maze to solve. You may move right, left, up, or down, so long as a clock hand does not oppose your motion. If you cannot cross through a clock more than once, what are the longest path and the shortest path through the maze? What is the best strategy to solve this?

13. The Upsilon Configuration

Figure F13.1 shows one possible solution. It misses only five blocks. Is it possible to do better?

F13.1. Solution to the Upsilon Configuration. (Illustration by Brian Mansfield.)

14. Bone Toss

Let L be the length of the stick or bone – whichever object the alien chooses to use. For enhanced creepiness, we can assume a bone is used. Let O be the center of the circle, and let P be the point where one endpoint of the bone lies on the circle (Figure F14.1). Choose a point Q on the circle so that $PQ = L$ and a is the measure of angle OPQ. We want to know the probability that the free end of the stick is in the circle. It turns out that the probability is $2a/2\pi$ (or $2a/360$, if the angle is measured in degrees). Why, you ask? Imagine one end of the bone fixed to the circle and the other end of the bone sweeping out a possible 360 degrees or 2π radians. Of this sweep, $2a$ of those radians lie within the original circle.

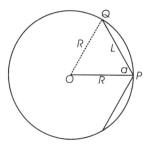

F14.1. Solution to the bone-toss problem.

The next step is to determine the value of *a*. We can use the handy *law of cosines,* along with the variables depicted in Figure F14.1, to compute the value for the angle *a* given the length of all threes sides of triangle *OPQ:*

$$R^2 = L^2 + R^2 - 2RL\cos(a)$$

The law of cosines is often used for calculating the length of one side of a triangle when the angle opposite and the other two sides are known. It can be also be used to find the angles of a triangle when the lengths of all three sides are known. (You'll soon see a way to avoid this formula for the first problem.)

In our first problem, the bone's length is the same as the circle radius, so $L = R$. *OPQ* turns out to be an equilateral triangle, and $OP = PQ = R$. Applying the law of cosines, we get $\cos(a) = 1/2$, and $a = \pi/3$. The probability is therefore $2\pi/3$ divided by 2π, or $1/3$. If you throw this bone and one end lies on the circle, you have a 33% chance that the "free" end of the bone is inside the alien's circle. Dorothy can tell the alien 33%.

Of course, for this case we didn't really need to use the law of cosines. If you had just recognized the existence of the equilateral triangle, you would have seen that $a = 60$ degrees or $\pi/3$ radians.

Throwing the longer bone of length $L = 2R$ results in $\cos(a) = 1$, and $a = 0$. Therefore, if we model the bone as a line segment of length $2R$ (= diameter of the circle), the probability of one end landing exactly *on* and the other *inside* the circle is 0. For the smaller bone of length $R/2$, we obtain $\cos(a) = 1/4$, and the probability is the inverse cosine of $(1/4)$ divided by π, or 0.4196. Note that the smaller bone has a greater chance of having its free end inside the circle than does the bone of length R.

The law of cosines is a wonderful formula, useful in a variety of practical applications. For example, it has applications in a variety of physics problems that use vector quantities, such as when we wish to find the difference between two vectors representing objects in a glancing collision (Figure F14.2).

F14.2. The law of cosines applied to vectors: $v_{final} = v_{initial} + \Delta v$ **or** $\Delta v^2 = v_i^2 + v_f^2 - 2v_iv_f\cos(a)$.

Here is another practical example to help illustrate situations in which the law of cosines provides an efficient means for solving a problem. Assume that in order for Dorothy to travel to the Oz facility from where she now stands she has to get on her bike and go 10 km directly east, then go 5 km directly northeast (45° north of east) (Figure F14.3). The aliens are trying to decide whether they should build a yellow-brick road that goes straight from where Dorothy stands to Oz. How many kilometers of travel will Dorothy save on each trip if this road is built?

F14.3. Usefulness of the law of cosines.

To solve this problem, we draw a schematic figure labeling Dorothy's current path to Oz. Because we know the values for a, b, and the angle between them, we can use the law of cosines to calculate the length of the yellow-brick road:

$$\begin{aligned}
c^2 &= a^2 + b^2 - 2ab\cos(\theta) \\
&= (10\text{ km})^2 + (5\text{ km})^2 - 2(10\text{ km})(5\text{ km})\cos(135°) \\
&= 100\text{ km}^2 + 25\text{ km}^2 - 100 \times (-0.7071)\text{ km}^2 \\
&= 125\text{ km}^2 + 70.71\text{ km}^2 \\
&= 195.71\text{ km}^2
\end{aligned}$$

Taking the square root of 195.71 km² we get the approximate answer $c = 14$ km. Thus, Dorothy will save about 1 km using the alternative path, because otherwise she would be traveling 15 km.

15. Animal Farm Courthouse

Here is one possible path, starting at 1, and passing through 24 cells and repeating

17	18	19	20	1	2
16	23	22	21	4	3
15	24	11	10	5	6
14	13	12	9	8	7

Are there other solutions?

16. The Omega Sphere

In theory, Dorothy should be able to find such a plane. It certainly exists.

Consider all the planes determined by triplets of points in the Omega Sphere. Next, pick a new line not wholly contained in one of these planes and outside of the sphere. Imagine a plane through this new line and to the left of the two billion points. Imagine rotating this plane about the new line and toward the right of the points. You can rotate this plane until it has passed through exactly one billion points. At this location, the plane divides the sphere into one billion points on each side of the plane.

Here is another way to look at this problem. The Omega Sphere contains a large but finite number of points. Imagine every possible line connecting any two of these points. Although there are a lot of lines, there is still a finite number of such lines. This means there must exist a plane such that no plane parallel to this plane contains any of these lines. Continuing with the same logic as before, if we start with a plane parallel to this plane and outside of the glowing sphere and then move the plane through the sphere, we will encounter the points one by one. We cannot encounter two of the points simultaneously because the plane cannot contain any of the line segments. Hence there will be a small "gap" between the billionth and the billion and first points, and any plane in this interval will be the dividing plane Dorothy seeks.

One of my colleagues asks a devious set of follow-up questions. Can you create a horizontal plane that divides the points equally? Remember that these are random points, so the answer is yes with probability 1. Now, imagine the usual x, y, and z axes in three-space with the positive z-axis pointing upward. Construct a vertical plane parallel to the x–z plane and parallel to the y–z plane such that the points are divided equally. Finally, look at the point where these three planes meet. Does every plane through this point divide the points equally? If you can't be sure, what is the probability that every plane through this point divides the points equally?

While on the topic of geometry, I cannot help but list my four favorite, little-known geometrical gems dealing with collections of points. More particularly, the following facts related to a collection of distances between points on a plane.

Visualize a collection of n ants ($n \geq 3$) stuck to a glass plane.

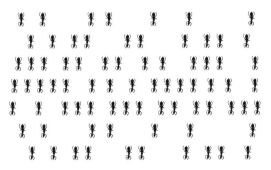

No matter where the ants might be positioned on the plane:

- The number of different distances produced must be at least $\sqrt{n - 3/4}$ – 1/2.
- The smallest distance produced cannot occur more often than $3n - 6$ times.
- The greatest distance produced cannot occur more often than n times.
- No distance produced can occur more than $n^{3/2}/\sqrt{2} + n/4$ times.

These four facts come from Ross Honsberger, *Mathematical Gems III* (New York: Mathematical Association of America, 1985), 36.

17. Leg-Bone Shatter Produces Triangle

A triangle can be formed only if the breaks are on opposite sides of the mid-point of the bone. If this is difficult for you to see, make some sketches. If the breaks happen to be on the same side of the midpoint, the two small pieces won't be able to make contact with one another when forming the triangle. After one break is created, the odds of creating a break on the opposite side of the bone's midpoint are 1 in 2. We have another restriction to consider. In order to form a triangle, the breaks must be less than half the bone's length apart. Again, sketch some diagrams if this is difficult to visualize. For this second condition, the probability is also 1/2. To determine the final combined probability, the two probabilities are multiplied together to obtain an answer of 1/4. Therefore, the chances of the Bone Being producing a triangle are 1 in 4. (Some of you may recall the *triangle-inequality theorem* that states that the sum of the lengths of any two sides of a triangle is greater than the length of the third side.)

Dr. Oz does not have an answer for the second part of the problem, and he welcomes reader feedback. Dr. Robert Stong suggests that there is no useful solution for the expected value of ratio of the smallest piece to the longest piece for reasons similar to those mentioned in the Further Exploring section for Chapter 8. The ratio is not a bounded function. However, it may be possible to derive mathematically an expression for the expected value of the ratio of the smallest piece to the longest piece.

18. Z-Bar Ranch

The following is one possible solution. (Cut the paper, or draw a line, so that all the symbols that fall in the gray squares are on one side). What is the most efficient way of solving this kind of problem?

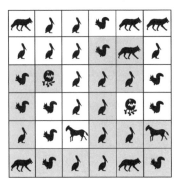

19. The Mystery of Phasers

Dr. Oz looks at the first problem this way. The chance of ship A hitting with its first shot is 50% or 0.5. Next we give B a chance. Fifty percent of the time B kills A. What is the chance that B misses and that A then shoots and kills B on A's second shot of the contest? We must find the compound probability of A first missing (0.5), followed by B missing, followed by A hitting. This is $0.5 \times 0.5 \times 0.5$. In the long run, we can gradually develop a series for the chance of A's winning:

$0.5 +$	A wins
$0.5 \times 0.5 \times 0.5 +$	A wins
$0.5 \times 0.5 \times 0.5 \times 0.5 \times 0.5 +$	A wins
$0.5 \times 0.5 \times 0.5 \times 0.5 \times 0.5 \times 0.5 \times 0.5 + \ldots + $	A wins

This long series becomes $0.5 \times (1 + 1/4 + 1/4^2 + 1/4^3 + \ldots) = 0.5 \times (4/3) = 2/3$. The value of 4/3 comes from the simple rule that tells us to what value a geometric series converges. Let a be any real number. Then the series

$$\sum_{n=0}^{\infty} a^n = 1 + a^2 + a^3 + a^4 + \ldots$$

is called a geometric series. In our case, $a = 1/4$. If $|a| < 1$, the geometric series converges. If $|a| \geq 1$, the geometric series diverges. If the geometric series converges, then it converges to

$$\sum_{n=0}^{\infty} a^n = \frac{1}{(1 - a)}$$

Because $a = 1/4$, all the terms of our series add up to 4/3.

Some colleagues have suggested that the problem can be examined from another angle. Of those contests that end on the first round, 2/3 are won by the first player. Of those contests that make it to the second round (second shot by each player) and end on that round, 2/3 are won by the first player.

Of those matches that end on the third round, 2/3 are won by the first player. And so on. Thus, of all matches, 2/3 are won by the first player.

Here's another way to look at the first problem. Let p be the probability of hitting your opponent. Let P be the probability of A surviving. (Remember, A is the ship that fires first.) This means that A will hit B with probability P and survive. With probability $(1 - p)^2$, both A and B will miss their first shots, leaving us in the same situation as at the beginning of the problem. This leads to the conclusion that $P = p + P(1 - p)^2$ or $P = p/(1 - (1 - p)^2)$. For $p = 1/2$, we find that $P = 2/3$ (0.67). For $p = 1/10$, we have $P = 100/190 = 0.53$. It's interesting to note that as both ships' accuracy decreases from 50% to 10%, ship A's chances of survival decreases. If your phasers are very inaccurate and the probability of hitting your opponent is a mere 1 out or a 100 or 0.01, then P is close to 0.5. This implies that as accuracy decreases there comes a point where there is essentially a 50–50 chance of A surviving. If the chance of hitting on each shot is very high, it becomes very likely that A will kill B with the first shot. Dorothy should therefore hope that her ship's phasers are as accurate as possible, even if ship B has this same higher level of accuracy.

One way to derive the aforementioned probability $p/(1 - (1 - p)^2)$ follows. In general, ship A can survive if it hits B right away (probability p), or ship A misses, B misses, and then A hits (giving a probability of $p \times (1 - p)^2$). If A and B miss twice first, and then A hits, we have a probability $p \times (1 - p)^4$, and so forth. The general expression is $p \times (1 - p)^{2n}$. The sum of the geometric series $p \times (1 - p)^{2n}$ is $p/[1 - (1 - p)^2]$ (first term divided by one minus the ratio), and this can be rewritten as $1/(2 - p)$.

In the Dreaded Fibonacci Gambit, let $F(n)$ be the nth Fibonacci term ($F(1) = 1$, $F(2) = 1$, $F(3) = 2$, $F(4) = 3$, ...). For a general probability p, ship A can survive if it hits B right away (probability p), or it misses, B misses, and then A hits. Here, in its second turn, A has two shots ($F(3) = 2$) and a probability of $[1 - (1 - p)^2](1 - p)^2$. If A and B both miss in the first two rounds (i.e., seven misses in all) and then A hits, we have five turns total with a probability of $[1 - (1 - p)^5](1 - p)^7$. We may continue this line of thinking further. My colleagues Ilan Mayer and David T. Blackston note that the general expression is $[1 - (1 - p)^{F(2n-1)}](1 - p)^{F(2n)-1}$. The sum of the series can be solved numerically. For $p = 0.5$ (50%) the solution is about 69.5%, which means that the first ship A has a 69% chance of winning the Fibonacci battle. For $p = 0.1$ (10%) ship A has a 54% chance.

For those of you who would like to experiment further with Fibonacci numbers, the following is BASIC code for computing the first 40 numbers in the Fibonacci sequence.

```
10  REM Compute Fibonacci Numbers
20  DIM F(40)
30  F(1)=1
40  F(2)=1
50  FOR N=1 TO 38
60     F(N+2)=F(N+1)+F(N)
70  NEXT N
80  FOR N=1 TO 40
90  PRINT F(N)
100 NEXT N
110 END
```

20. Salty Logic

Here is one solution, $33 \times 2 \div 11 = 6$. There are others. Can you find them?

	3	×	2
3	🐙	🐙	÷
E	🐙	🐙	1
6	=	1	

21. Where Are the Composites?

How in the world should Dorothy go about attacking this problem? It certainly seems difficult to find 10,000 consecutive nonprime numbers using pencil and paper! What do we know about the distribution of primes?

- One way of finding the prime numbers is to use the ancient *Sieve of Eratosthenes.* First, you make a list of positive numbers, and then all the multiples of 2 are eliminated, starting at 4. Then all the multiples of 3 are eliminated, starting at 6; the process is repeated until all possible eliminations have taken place. (A modern computerized version of the Sieve is one of the traditional ways of evaluating and comparing computers because the process is lengthy and computationally intensive.)
- The *prime-number theorem* states that the number of primes less than n can be approximated by $n/(\ln n)$. This theorem was first conjectured by Carl Friedrich Gauss in the early nineteenth century and was proved (independently) by Jacques Hadamard and Charles de la Vallée Poussin in 1896. Their proofs relied on complex analysis, and, at the time, no one thought a more simplified proof could be constructed. The great mathematical event of 1949 was another proof of the prime-number theorem, given by Atle Selberg and Paul Erdös. (Incidentally, the theorem can be used to derive a related theorem: For each number greater than 1, there is always at least one prime number between it and its double. Also, one

can show from the prime-number theorem that the average "gap" between primes less than n is $\ln(n)$. [If you examine the first few primes: 2, 3, 5, 7, 11, and 13, you'll notice the differences between successive primes varies as: 1, 2, 2, 4, 2 . . .])

- For centuries, mathematicians have tried to explain the underlying pattern behind the primes. Perhaps no pattern exists. Certain prime numbers occur in pairs, separated by a single even number. Here are some of these pairs, called *twin primes:* $(3, 5)$ $(5, 7)$ $(11, 13)$ $(17, 19)$ $(29, 31)$. A longstanding conjecture of mathematics holds that there is an infinite number of twin primes. So far, no proof or disproof has been formulated. (Notice that twin primes differ by 2, which is as close as primes can be to each other. If they differed by 1, one of the numbers would have to be even and therefore divisible by 2.) Will we ever develop a convenient formula for generating all prime numbers, or one that exactly counts the number of primes up to a particular large number?

Can we use this odd assortment of facts to help us solve Dorothy's problem? Perhaps, but there's an easier way for Dorothy to come up with 10,000 composite numbers.

One "easy" answer is $10{,}001! + 2$, $10{,}001! + 3$, . . . $10{,}001! + 10{,}001$, where "!" is the factorial symbol. (For example, $5! = 5 \times 4 \times 3 \times 2 \times 1$.) The sequence is nonprime (composite) because $n! + A$ is divisible by A when $A > 1$ and ($A \le n$). Let's examine an example of $n! + A$, where $n = 5$ and $A = 3$, which yields $(5 \times 4 \times 3 \times 2 \times 1) + 3$. The factorial part of this number must be divisible by 3 because it contains all the factors from 1 to n. The second part of the sum is obviously divisible by 3. Similarly, if $A = 4$, we find that $120 + 4$ is divisible by 4.

Although our "easy" answer is a correct one that should satisfy the alien, it is not a string of numbers with the 10,000 *smallest* values that would answer the problem. I am unaware of an easy way to find the smallest string of 10,000 composite numbers.

While on the topic of prime numbers, I can't help mentioning that cicadas spend 7, 13, or 17 years beneath the ground before emerging as adults for the final weeks of their lives. Significant research is being conducted on how evolutionary forces might have led to these prime-number life cycles. For example, recent work with mathematical simulations shows that prime-number periodicities allow cicadas to escape from predators. Isn't it amazing that number theory plays a key role in understanding zoology? For more information, see A. M. S., "Biological model generates prime numbers," *Science* 293 (5528) (July 13, 2001): 177.

22. Brain Trip

Figure F22.1 shows one solution. Is it the only one? How long did it take you to solve this, and what is the best strategy for finding a solution?

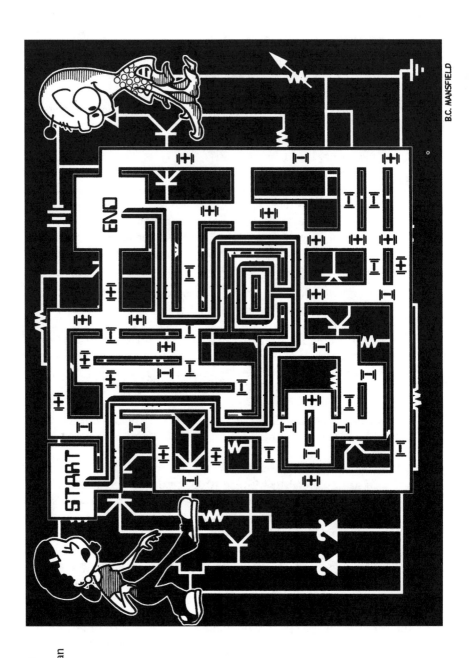

B.C. MANSFIELD

F22.1. Solution to Brain Trip. (Illustration by Brian Mansfield)

23. The Gaps of Omicron

Dr. Oz believes that the following is an expression for all rational solutions:

$$\alpha = (1 + 1/k)^k; \qquad \beta = (1 + 1/k)^{k+1}$$

for $k = \pm 1, \pm 2, \pm 3, \ldots$, but Dr. Oz is not sure what a plot looks like. Would initial attempts at drawing the graph contain gaps because Oz is interested only in rational solutions?

Here is how we arrive at the solutions for α and β. Let's start with the initial formula

$$\alpha^\beta = \beta^\alpha$$

If we let $\beta = r\alpha$, we get $\alpha^{r\alpha} = (r\alpha)^\alpha$. Next, $\alpha^r = r\alpha$, or $\alpha^{r-1} = r$, or $a = r^{1/(r-1)}$. Now, Dr. Oz said from the start that we are interested in rational numbers. Assuming r is rational, α will be rational whenever $1/(r-1)$ is an integer, that is, when $r = 1 + 1/k$. Thus, we find $\alpha = (1 + 1/k)^k$; $\beta = (1 + 1/k)^{k+1}$ for $k = \pm 1, \pm 2, \pm 3, \ldots$.

Darrell Plank notes that it should be possible to create a plot using the software package Mathematica, for which the relevant command would be:

ListPlot[Table[{(1. + 1./k)^k, (1. + 1./k)^(k + 1)}, {k, −100, 100}]]

What would this plot look like? What happens as larger values for k are used?

Dr. Oz guesses that the plot looks somewhat like a hyperbola, passing through $(2, 4)$ and $(4, 2)$. He welcomes plots from readers. A similar problem, without graphical concerns, is discussed in Angela Dunn, *Mathematical Bafflers* (New York: Dover, 1980), 213.

24. Hutchinson Puzzle

Here is one solution. How many other solutions can you find?

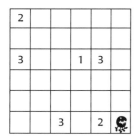

25. Flint Hills Series

The Flint Hills series $S(N)$ is an unusual one. If you plot $S(N)$ as a function of N, it appears to converge nicely for the first 354 terms. Here is a table of values, starting at $N = 22$, so that you can see the rise, and smooth leveling off, to a value near 4.8.

N	S	N	S
22	4.75410	307	4.80686
23	4.75422	308	4.80686
24	4.75430	309	4.80686
25	4.75796	310	4.80686
26	4.75806	311	4.80687
27	4.75811	312	4.80687
28	4.75873	313	4.80687

Apparent convergence to a value close to 4.80687

Dr. Oz computed these values using the following recipe:

```
ALGORITHM: How to Compute Strange Series
s=0
DO n = 1 to 400
  olds=s
  s = s + 1./(n**3 * (sin(n))**2)
  PrintValueFor(n, s)
  if ((s−olds) > 3) then Print("I have found a jump")
END
```

The series seems pretty tame at this point. In fact, just by looking at the values, I would have guessed that the series was converging to a value near 4.8. However, at $N = 355$, the series' values suddenly jump up to 29.4! Holy mackerel! Dr. Oz nearly lost his teeth when he saw this unexpected jump. This is a fine example to demonstrate to students the danger of looking at graphs and tables of numbers in order to assess convergence.

Why does the seemingly well-mannered behavior suddenly skyrocket at $N = 355$? The reason is simple when viewed in hindsight. First, recall that $\sin(N\pi) = 0$. Since 355 is almost a precise multiple of π ($355/113 = 3.14159$ is an excellent approximation), the values for $S(N)$ jump very abruptly at this point. Further jumps might occur later, whenever an excellent rational approximation to π is encountered.

Can you find other jumps? Dorothy has examined the first 100,000 terms and does not find a large jump other than at $N = 355$. What would an infinitely patient mountain climber find as he "walked" along this series for an infinite number of miles? Mathematicians have worked on the frequency of rational approximations to π, but their knowledge is not yet sufficient to answer whether this series converges or not. Figure F25.1 shows a plot of those values of N between 0 and 10,000 that are almost multiples of π. Stated more mathematically, the figure is a plot of N for $|N/k - \pi| < \varepsilon$, where ε is 0.0001. The points at $\varepsilon \sim 0$ are located at multiples of 355. (They appear to be on the zero axis due to the resolution of the graph.)

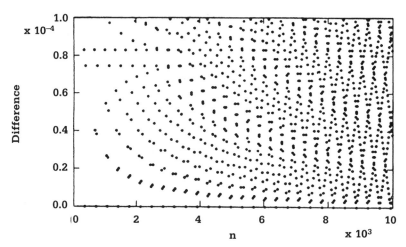

F25.1. Close approximations to pi.

Here is recipe for computing this graph.

```
ALGORITHM: How to Create the Pi Dot Map
pi = 3.1415926
DO k = 1 to 10000
  DO n = k to 10000
    ratio = n/k
    diff = abs(ratio – pi)
    if (diff < .0001) then PlotPointAt(n,diff)
  END
END
```

Dr. Oz wishes to thank Ross McPhedran, Department of Theoretical Physics, The University of Sydney, Sydney NSW 2006 Australia, for useful comments and for bringing the Flint Hills series to Dr. Oz's attention. Ross attributes this equation to the French physicist Professor Roger Petit, Laboratoire d'Optique Électromagnetique, Faculté des Sciences et techniques de St. Jérome, Marseille 13397, France.

Darrell Plank notes that it is possible to study this series graphically using the software package Mathematica with the following command:

ListPlot[FoldList[(#1 + 1./(#2^3 (Sin[#2]^2))) &, 0, Range[1000]]]).

Using a 700MHz Intel Celeron processor and the QBASIC language, Jason Earls has recently summed the first 15,000,000 terms of the series and obtained a value of 30.31454606891396. He also is currently working on producing a list of values for those n at which point the series jumps (e.g., 355, 710). Jason has also summed the first 2,631,403 terms of

$$S(N) = \sum_{n=1}^{N} \frac{1}{n^3 \cos^2 n}$$

Cookson Hills series

(Jason named the series after the Cookson Hills, which are 20 miles from his ancestral home in Eastern Oklahoma.) This produces a similar series to the Flint Hills series, and it exhibits two major jumps in the first few dozen terms, after which the series appears to settle to a value near 42.99523402763187.

26. Wacky Tiles

The answer is:

When completed, the square reads the same both down and across.

Here is another puzzle for you to consider. Fill in the empty tile with the correct missing symbols.

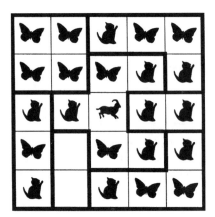

27. Toto Clone Puzzle

Three of the clones are standing on the points of an equilateral triangle △. The fourth Toto is on a small rise in the meadow at the center of the triangle. The four Toto clones thus form the four corners of an equal-sided triangular pyramid (tetrahedron).

28. Legion's Number

Let's first tackle the question of Legion's number of the second kind. In some sense, that's actually easier than solving the first problem. It is not as difficult as it appears at first glance to find the last ten digits of

$$\text{M} = 666!^{666!}$$

Here is a list of the first 40 factorials.

```
 1! = 1
 2! = 2
 3! = 6
 4! = 24
 5! = 120
 6! = 720
 7! = 5040
 8! = 40,320
 9! = 362,880
10! = 3,628,800
11! = 39,916,800
12! = 479,001,600
13! = 6,227,020,800
14! = 87,178,291,200
15! = 1,307,674,368,000
16! = 20,922,789,888,000
17! = 355,687,428,096,000
18! = 6,402,373,705,728,000
19! = 121,645,100,408,832,000
20! = 2,432,902,008,176,640,000
21! = 51,090,942,171,709,440,000
22! = 1,124,000,727,777,607,680,000
23! = 25,852,016,738,884,976,640,000
24! = 620,448,401,733,239,439,360,000
25! = 15,511,210,043,330,985,984,000,000
26! = 403,291,461,126,605,635,584,000,000
27! = 10,888,869,450,418,352,160,768,000,000
28! = 304,888,344,611,713,860,501,504,000,000
29! = 8,841,761,993,739,701,954,543,616,000,000
30! = 265,252,859,812,191,058,636,308,480,000,000
31! = 8,222,838,654,177,922,817,725,562,880,000,000
32! = 263,130,836,933,693,530,167,218,012,160,000,000
33! = 8,683,317,618,811,886,495,518,194,401,280,000,000
34! = 295,232,799,039,604,140,847,618,609,643,520,000,000
35! = 10,333,147,966,386,144,929,666,651,337,523,200,000,000
36! = 371,993,326,789,901,217,467,999,448,150,835,200,000,000
37! = 13,763,753,091,226,345,046,315,979,581,580,902,400,000,000
38! = 523,022,617,466,601,111,760,007,224,100,074,291,200,000,000
39! = 20,397,882,081,197,443,358,640,281,739,902,897,356,800,000,000
40! = 815,915,283,247,897,734,345,611,269,596,115,894,272,000,000,000
```

Isn't that an impressive pile of numbers? Notice how the number of trailing zero digits grows. You can see that by 40! There are nine trailing zeros. This implies that the last 10 digits of 666! are 0 and thus that many more than the last 10 digits of И are also 0. The multitude of trailing zeros arises from all the factors of 5 and 2 that get multiplied together. More precisely, the last 10 digits of 666! are 0 because 5^{10} and 2^{10} both divide 666! This means that at least the last 10 digits of И are also 0. Note that n! will contain factors for every integer ending in 0 or 5 up to n, and each of those will produce at least one extra 0 at the end. Anything else with an extra power of 5 (e.g., 25) will produce another 0. This means 666! will have well over 100 zeros.

What can we know about the last ten digits of Legion's number of the first kind?

$$\text{Ͷ} = 666^{666}$$

Dorothy might have trouble with this without a personal computer. Today we can actually compute all 1881 digits using a high-precision calculator such as the Unix "bc" calculator. The last ten digits of Ͷ are 0,880,598,016.

Researcher Jim Gillogly has written a program using another arbitrary precision software package called GMP (Gnu Multi-Precision) that explores both kinds of Legion's numbers. For example, the program calculates the last 10 digits of Ͷ, verifying the previously mentioned "bc" result. Next, the program strips off the zeros from the end of 666!, then raises that number to the 666! power modulo 10,000,000,000 to obtain the last 10 digits of И before the trailing zeros. The results show that Ͷ terminates with 880598016. И terminates in 1787109376 followed by 165 zeros. Here is the program Jim Gillogly used for performing the massive computation:

```
/* Legion's Numbers: last 10 digits of 666^666 and 666! ^ 666!
 * Jim Gillogly, Dec 2000 */
#include <stdio.h>
#include <gmp.h>
main(int argc, char **argv)
{
    mpz_t legion; /* Allocate space for the multi-precision integers. */
    mpz_t result;   mpz_t mod;   mpz_t fact;   mpz_t remainder;
    mpz_t base;   int i;
    mpz_init(legion);   /* Initialize multi-precision variables. */
    Mpz_init(result);   mpz_init(mod);   Mpz_init(fact);   mpz_init(base);
    mpz_init(remainder);
    mpz_set_str(mod, "10000000000", 0);   /* Last 10 digits. */
    mpz_set_str(legion, "666", 0);
    mpz_powm(legion, legion, legion, mod); /* 666^666 mod 10^10 */
    printf("Last 10 digits of 666^666: ");
    mpz_out_str(stdout, 10, legion); printf("\n"); /* Print the value. */
    mpz_set_ui(fact, 2);   /* Calculate 666! */
```

```
for (i = 3; i <= 666; i++)
    mpz_mul_ui(fact, fact, i); printf("666! = ");
mpz_out_str(stdout, 10, fact); printf("\n"); /* Print 666! */
/* 666! ends in a whole bunch of 0's: you get another one whenever
 * you multiply in another factor ending in 0 or 5.
 * Just for fun, let's strip them all off (counting them), and
 * find out what the last digits are <before> the zeroes. */
mpz_set(base, fact);   /* Keep the full # for exponent. */
for (i = 0; i < 500; i++)   /* Under 200 zeroes, actually. */
{
    mpz_tdiv_qr_ui(result, remainder, base, 10);
    if (mpz_cmp_ui(remainder, 0) != 0)
        break;   /* Finished stripping zeroes. */
    mpz_set(base, result);   /* Strip off that zero. */
}
printf("We stripped %d zeroes from the end of 666!\n", i);
mpz_powm(result, base, fact, mod);   /* Raise it to the 666! power. */
printf("Last n digits of (666!)^(666!) (not counting 0's): ");
mpz_out_str(stdout, 10, result);   printf("\n");
return 0;
}
```

The program output follows:

Last 10 digits of N: 880598016
666!=10106320568407814933908227081298764517575823983241454113404
20807357413802103697022989202806801491012040989802203557527 0
39339704057130729302834542423840165856428740661530297972410 6
82828699397176884342513509493787480774903493389255262878341 7
61883261899426484944657161693131380311117619573051526423320 3
89641805410816067607893067483259816815364609828668662748110 3
85603657973284604842078094141556427708745345100598829488472 5
05949071967727270911965060885209294340665506480226426083357 9
01503097781140832497013738079112777615719116203317542199999 4
89227144752667085796752482688850461263732284539176142365823 9
73696764537603278769322286708855475069835681643710846140569 7
69330065775414413083501043659572299454446517242824002140555 1
40464296291001901438414675730552964914569269734038500764140 5
51143642836128613304734147348086095123859660926788460671181 4
69216252213374650499557831741950594827147225699896414088694 2
51261045196672567495532228826719381606116974003112642111561 3
32573503212960729711781993903877416394381718464765527575014 2
52129040283236963922624344456975024058167368431809068544577 2
58472983797943781807264821360865009874936976105696120379126 53
63665664696802245199962040041544438210327210476982203348458 5
```
```

96093079296569561267409473914124132102055811493736199668788534872321705360511305248710796441479213354542583576076596250213454667968837996023273163069094700429467106663925419581193136339860545658673623955231932399404809404108767232000

We stripped 165 zeroes from the end of 666!

Last n digits of (666!)^(666!) (not counting 0's): 1787109376

Darrell Plank used the Mathematica software package and logarithms to attack an even more difficult problem: What are the *first* 10 digits of Legion's number of the second kind? First he computed Legion's number of the first kind

$$666^{666} = 2.7154175928871285582608746 \times 10^{1880}$$

Thus, the 10 digits are 2715417592. Using base 10, we find log(666!) = 1593.00459306976630675482696146. If we call this L, then

$$
\begin{aligned}
&\log(666!^{666!}) \\
&= 666! \times L = (10^L) \times L \\
&= 1.01063205684078149339082271 \times 10^{1593} \\
&\quad \times 1593.00459306976630675482696146 \\
&= 1.60994150845091005477479540 \times 10^{1596}
\end{aligned}
$$

Let $M = 1.60994150845091005477479540$. This gives us $M \times 10^{1596}$. Taking 10 to this power gives

$$(10^M) \times 10^{10^{1596}}$$

Obviously the factor on the right is a large but integral power of 10. The left gives $10^M = 40.732541480224810426094063$, so the first 10 digits of $666!^{666!}$ should be 4073254148. Isn't that a dandy analysis? Plank notes that the way to obtain log(666!) isn't obvious. One can note that log(666!) = log(1 × 2 × 3 × . . . × 665 × 666) = log(1) + log(2) + . . . + log(666).

Readers interested in factorials may wish to conduct research on *factorial primes* of the form "factorial plus one" (that is, $n! + 1$) and "factorial minus one" (that is, $n! - 1$). Various researchers have discovered that the form $n! + 1$ is prime for $n = 1, 2, 3, 11, 27, 37, 41, 73, 77, 116, 154, 320, 340, 399, 427, 872,$ 1477, and 6380 (which corresponds to a number with 21,507 digits). (See Borning 1972; Templer 1980; Buhler, Crandall, & Penk 1982; and Caldwell 1995.) The form $n! - 1$ is prime for $n = 3, 4, 6, 7, 12, 14, 30, 32, 33, 38, 94, 166,$ 324, 379, 469, 546, 974, 1963, 3507, 3610, and 6917 (23,560 digits). Both forms have been tested to $n = 10,000$ (Caldwell & Gallot, 2002). These and related numbers are discussed by Chris K. Caldwell, "Primorial and factorial primes," http://www.utm.edu/research/primes/lists/top20/PrimorialFactorial.html.

You can learn more about Dr. Chris K. Caldwell, who is a professor at the Department of Mathematics and Statistics, University of Tennessee at Martin, at his Web site, http://www.utm.edu/~caldwell/. His current academic interests include prime-number theory and the use of the computers to teach mathematics.

Note that one of my favorite numbers is called the Beast:

25.8069758	0112788031	5188420605	1491408960	8260667187
2206858524	1369237122	8080398905	1038349992	6720968861
9318855007	5761727345	9109615963	7558843433	2985885744
0382590747	9275606091	5875182845	9438603101	2881616726
4637737512	8227234943	1038556483	2857611197	8689357746
9533562989	5358928362	1920680996	4202011090	5005852090
50268385				

Can anyone guess why this number is of particular interest?

29. The Problem of the Tombs

For the first problem, if we assume that the people diffuse randomly like particles in a gas, the number of people in each tomb is proportional to the area of the tomb. Therefore, most people will be in the last tomb, *J*.

We should approach the second problem by thinking about human psychology. For example, if people thought that there was chance of escape, retrieval, or visits, a higher density might remain in tomb *A*, where the people were originally added. Other factors might be considered; for example, the acoustical properties of the different chambers might provide different kinds of echoes, and the quietest chamber might be most desirable. Also, we might wish to consider how chamber size affects oxygen movement, humidity, diffusion of body odors, and an individual's sense for adventure or desire for solitude. In which chamber would you prefer to live? Using your creative brains, Dr. Oz bets you could come up with other valid solutions.

Roland Tomlinson says about the second problem:

The strongest, most forceful humans would take control and claim the smallest tomb for themselves because there is "prestige" value for being a member of a small set, namely the set of people in the smallest tomb. The weaker people would be relegated to the larger tombs. The furthest end of the largest tomb would be the obvious place for waste disposal, so I imagine the aforementioned "rulers" would recruit some minions to trek the miles required to dispose of waste.

30. Mr. Plex's Tiles

The answer is 𝕍𝕍𝕍 𝕍𝕍𝕍 𝕍𝕍𝕍, because each row has four snakes and two Mr. Plexes. Can you justify another solution? How long did this take you to solve?

Another area of research include *primorial primes*. Let $p\#$ (which is read as "p-primorial") be the product of the primes less than or equal to p so that

$$3\# = 2 \times 3 = 6,$$
$$5\# = 2 \times 3 \times 5 = 30, \quad \text{and}$$
$$13\# = 2 \times 3 \times 5 \times 7 \times 11 \times 13 = 30,030$$

Primorial primes also have two forms: "primorial plus one" ($p\# + 1$) and "primorial minus one" ($p\# - 1$). $p\# + 1$ is prime for the primes $p = 2, 3, 5, 7, 11,$ 31, 379, 1019, 1021, 2657, 3229, 4547, 4787, 11,549, 13,649, 18,523, 23,801, 24,029, and 42,209 (which has 18,241 digits). (See Borning 1972; Templer 1980; Buhler, Crandall, & Penk 1982; and Caldwell 1995.) $p\# - 1$ is prime for primes $p = 3, 5, 11, 13, 41, 89, 317, 337, 991, 1873, 2053, 2377, 4093, 4297,$ 4583, 6569, 13,033, and 15,877 (6845 digits). Both forms have been tested for all primes $p < 100,000$ (Caldwell & Gallot 2002). (For more information on primorial and factorial primes, see Dubner 1987, 1989.)

Finally, Chris Caldwell and Harvey Dubner generalize the factorial primes by using the multifactorial functions:

$$n! = (n)(n-1)(n-2)\ldots(1)$$
$$n!! = (n)(n-2)(n-4)\ldots(1 \text{ or } 2)$$
$$n!!! = (n)(n-3)(n-6)\ldots(1, 2, \text{ or } 3)$$

For example, $7! = 5040$, $7!! = 105$, $7!!! = 28$, $7!!!! = 21$, and $7!!!!! = 14$. Multifactorial primes are primes of the forms $n!! +/- 1$, $n!!! +/- 1$, $n!!!! +/- 1$, and so on. (See Caldwell and Dubner 1993.) Here are references on primorial, factorial, and multifactorial primes supplied by Chris Caldwell:

- Borning, A. (1972). "Some results for $k! \pm 1$ and $2 \cdot 3 \cdot 5 \ldots p \pm 1$," *Mathematics of Computation* 26: 567–70.
- Buhler, J. P., R. E. Crandall, and M. A. Penk (1982). "Primes of the form $n! \pm 1$ and $2 \cdot 3 \cdot 5 \ldots p \pm 1$," *Mathematics of Computation* 38: 639–43. Corrigendum in *Mathematics of Computation* 40 (1983): 727.
- Caldwell, C. (1995). "On the primality of $n! \pm 1$ and $2 \cdot 3 \cdot 5 \ldots p \pm 1$," *Mathematics of Computation* 64(2): 889–90.
- Caldwell, C., and H. Dubner (1993/4). "Primorial, factorial and multifactorial primes," *Mathematical Spectrum* 26(1): 1–7.
- Caldwell, C., and Y. Gallot (2002). "On the primality of $n! \pm 1$ and $2 \times 3 \times 5 \times \ldots \times p \pm 1$," *Mathematics of Computation* 71(237): 441–8.
- Dubner, H. (1987). "Factorial and primorial primes," *Journal of Recreational Mathematics* 19(3): 197–203.
- Dubner, H. (1989). "A new primorial prime," *Journal of Recreational Mathematics* 21(4): 276.
- Templer, M. (1980). "On the primality of $k! + 1$ and $2 \times 3 \times 5 \times \ldots \times p + 1$," *Mathematics of Computation* 34: 303–4.

Here is another challenge for you. Fill in the empty tile with the correct missing symbols.

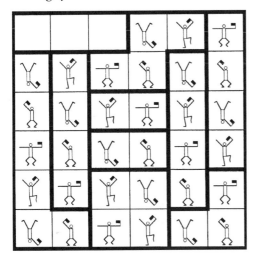

31. Phasers on Targets

Consider a target, schematically drawn as a rectangle here for simplicity of typesetting:

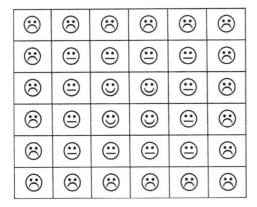

Target

The "goodness" level of a shot can be represented by ☺ (best shot), ☺ (intermediate shot), ☹ (worst shot). Although the target is represented with a 6-by-6 array of cells, it's really a continuous range of positions, and it's virtually impossible that two randomly fired shots will be identically close to the center. Therefore of Dorothy's three shots, one shot will be the best, another will be intermediate, and one will be the worst. There are six possible outcomes of the three shots. Let's list them.

	1	2	3	4	5	6
1st shot	☺	☺	😐	😐	☹	☹
2nd shot	😐	☹	☺	☹	☺	😐
3rd shot	☹	😐	☹	☺	😐	☺

Outcomes

For example, it's possible that the first shot is the best, the second shot is intermediate, and the third shot is worst. This situation is depicted in column 1. But it's equally possible that the best shot is first, the third shot is intermediate, and the second shot is worst. This is shown in column 2.

However, we already have two pieces of information that restrict the various possible outcomes. The first shot is not the worst shot, as evidenced by the words, "Darn, her second shot is even further from the center than the first." We also know that the second shot is not the best shot. This means we can eliminate the scenarios depicted by columns 3, 5, and 6. For example, column 3 shows the second shot as the best, but we know that this is not possible. We are left with scenarios 1, 2, and 4, and each appears equally likely.

Dorothy's mission was to determine the probability that her final shot is further from the center than her first. We can examine columns 1, 2, and 4. Of these, 1 and 2 have the third shot worse than the first. This means that the probability that Dorothy's last shot is further from the center than her first is 2/3. If you were a gambling person, this would be a good bet.

We may be able to represent the problem more generally. Again we assume that all the shots are independent and identically distributed. This is implied by Dr. Oz when he says, "Assume that your skill level stays constant." Also continue to assume that there is zero probability that two shots are exactly the same distance from the target. Suppose Dr. Oz fires a total of N shots at the target, and that $N>2$. My colleague David Karr has noted we can enumerate $N(N-1)$ cases or outcomes, each one specifying the ranks (in the list of shots ordered from best to worst) of just the first and last shots. For each of the N places in the ranking in which the first shot may fall, there are $N-1$ remaining places where the last shot may fall. In our three-shot puzzle, $N=3$, and the number of possibilities or outcomes was 6. This corresponds to the 6 columns in the previous table. Each of these cases is equally likely. Now suppose that of the $N-2$ shots between the first and the last, M of them are better than the first, and $N-M-2$ are worse than the first. This eliminates some of the cases enumerated, with the following cases remaining:

(a) First shot has rank $M+1$; last shot has one of the $N-M-1$ ranks worse than $M+1$ (i.e., there are $N-M-1$ subcases.)

(b) First shot has rank $M+2$; last shot has one of the $M+1$ ranks better than $M+2$ (i.e., there are $M+1$ subcases.)

Altogether there are N equally likely subcases, so the probability of case (b) is $(M+1)/N$. This means that, in general, we can arrive at the "Karr formula," which gives the probability that the last shot is better than the first, assuming that M shots are better than the first:

$$(M+1)/N$$

In our first puzzle, M was 0 and $N=3$, yielding a probability of 1/3 that the last shot is better than the first. If Dorothy shoots 2001 times, and 285 shots are better than the first one, what do we know about the 2002d shot? In this case, $N=2002$ and $M=285$, which gives the probability $286/2002 = 1/7$ that the last shot is better than the first. If Dorothy shoots 2002 times and one shot is better than the first, $N=2002$ and $M=1$, which gives only a 2/2002 chance that the last shot is better than the first. Do the odds that the last shot is better than the first decrease as the number of shots increases?

32. The Chamber of Death and Despair

Since Dr. Oz always points to a tunnel that is sticky, the odds greatly favor Dorothy if she switches to the remaining tunnel, which in this case is Tunnel 2. If you don't believe this suggestion, write a computer simulation. Alternatively, it may be helpful for you to imagine there were 100 tunnels, and Dr. Oz pointed to the 98 sticky ones. Now would you switch your choice of tunnels? (What assumptions are being made in this analysis?)

33. Zebra Irrationals

In Carl Sagan's science-fiction novel *Contact,* researchers search for patterns in the endless digits of π. As Sagan notes, the ancient Hebrews thought that π was exactly 3. The Greeks and Romans had no idea that the digits went on forever without repeating, something humans discovered only about 250 years ago. Without giving too much of the plot away, the researchers in the novel do eventually find a secret message hidden deep within the digits of π, which leads people to believe that the universe was made on purpose. Perhaps scientists would discover more messages if they looked even further into π. Sagan's characters conclude that, "In the fabric of space and in the nature of matter, as in a great work of art, there is, written small, the artist's signature. Standing over humans, gods, and demons, subsuming Caretakers and Tunnel builders, there is an intelligence that antedates the universe."

The square root of 2, like π, is an endless pattern of nonrepeating digits. When the square root of 2 was first proved to be irrational (i.e., it could not be expressed as the ratio of two integers like 7/5), a whole new area of mathematics was developed. The Pythagoreans, a mystical brotherhood based on the philosophical teachings of Pythagoras, discovered that the diagonal of a square with sides of length 1 is not a rational number. This was considered so shocking that those who knew about it were sworn to secrecy for fear that it might disrupt the fabric of society! It is often said that when Hippasus discovered that the ratio between the side and the diagonal of a rectangle cannot be expressed in integers, this shattered the Pythagorean world view. The problem caused an existential crisis in ancient Greek mathematics. The digits of 1.4142 . . . go on forever without any known pattern. Pythagoreans dubbed these irrational numbers *alogon,* or unutterable.

The construction of zebra numbers was prompted by the standard claim that the digits of an irrational number chosen at random would not be expected to display obvious patterns in the first one hundred digits. Some have said that if such a pattern were found, it would be irrefutable proof of the existence of either God or extraterrestrial intelligence. Obviously, zebra irrationals do indeed display spectacular patterns.

The particular zebra irrational displayed in this chapter was supplied by Robert Israel from the Department of Mathematics, University of British Columbia, Vancouver, B.C., Canada. Additional computation appears to reveal no more obvious patterns:

```
272727272727272727272727272727272727.272727272727272727272727272
727089696969696969696969696969696969696969696969696969696969696969
969696969082801346801346801346801346801346801346801346801346801346
0134680134676012928095772540216984661429105873550317994762439206883650982326573720746560252733092239265078771251610757783597289737642339133033120413092978250728593664121784835090733581813987734444519175599925485613642906738471438974411082321461491887297847754271123229107276254560713230631434157319438997012303659626786754085171783467549131094096574742591892186605841916803187303515235612642056212984646767094447604189085745997334005826903153349056181004751489365172251727071956119358210864920037337022498578713536668432027618877439815713532857035704684006128958095753478236319623793182127481126984568641966410467734106927885845644963942555395946653455423520325433044141070690578892528600963565039853372102383551160510486472004418239179360370026841516202413688777890867539436696628182006953258857566552910708650481935148640288493106982078...
```

Jason Earls from Blackwell, Oklahoma, is the world's most persistent zebra hunter, and he holds the world record for computing this zebra irrational to 20,000 digits. He has also computed $f(50)$ to show that with larger values of n the resulting zebra displays more persistent patterns:

$$f(50) = \sqrt{9/121 \times 100^{50} + (112 - 44 \times 50)/121}$$

The digits of $f(50)$ start with:

2

72727 27272 72727 27272 72727 27272 72727 27272 72727 2727.2
72727 27272 72727 27272 72727 27272 72727 27272 72727 26956
36363 63636 36363 63636 36363 63636 36363 63636 36363 63636
36363 63636 36363 63636 36363 63636 36363 63636 36363 45287
27272 72727 27272 72727 27272 72727 27272 72727 27272 72727
27272 72727 27272 72727 27272 72727 27272 72727 27251 44232
72727 27272 72727 27272 72727 27272 72727 27272 72727 27272
72727 27272 72727 27272 72727 27272 72727 27272 69640 95563
63636 36363 63636 36363 63636 36363 63636 36363 63636 36363
63636 36363 63636 36363 63636 36363 63636 36358 62418 46807
27272 72727 27272 72727 27272 72727 27272 72727 27272 72727
27272 72727 27272 72727 27272 72727 27272 71855 15358 89919
99999 99999 99999 99999 99999 99999 99999 99999 99999 99999
99999 99999 99999 99999 99999 99999 99998 41025 13993 62559
99999 99999 99999 99999 99999 99999 99999 99999 99999 99999
99999 99999 99999 99999 99999 99999 99700 33238 87798 42559
99999 99999 99999 99999 99999 99999 99999 99999 99999 99999
99999 99999 99999 99999 99999 99999 42064 26183 07695 61599
99999 99999 99999 99999 99999 99999 99999 99999 99999 99999
99999 99999 99999 99999 99999 99885 75072 43302 7758

Jason Earls also conducted experiments with the zebra irrational $f(95)$. The strange and beautiful patterns of repeating digits appear to persist for about 12,500 digits, and we wonder if the subsequent apparently "random" digits actually contain patterns discernible by statistical analysis.

$$f(95) = \sqrt{9/121 \times 100^{95} + (112 - 44 \times 95)/121}$$

The digits of $f(95)$ start with:

2

72727 27272 72727 27272 72727 27272 72727 27272 72727 27272
72727 27272 72727 27272 72727 27272 72727 27272 7272.7 27272
72727 27272 72727 27272 72727 27272 72727 27272 72727 27272
72727 27272 72727 27272 72727 27272 72727 26656 36363 63636
36363 63636 36363 63636 36363 63636 36363 63636 36363 63636
36363 63636 36363 63636 36363 63636 36363 63636 36363 63636
36363 63636 36363 63636 36363 63636 36363 63636 36363 63636
36363 63636 36363 63636 36362 93987 27272 72727 27272 72727
27272 72727 27272 72727 27272 72727 27272 72727 27272 72727
27272 72727 27272 72727 27272 72727 27272 72727 27272 72727

27272 72727 27272 72727 27272 72727 27272 72727 27272 72727
27272 72727 27115 32032 72727 27272 72727 27272 72727 27272
72727 27272 72727 27272 72727 27272 72727 27272 72727 27272
72727 27272 72727 27272 72727 27272 72727 27272 72727 27272
72727 27272 72727 27272 72727 27272 72727 27272 72727 27272
28259 81063 63636 36363 63636 36363 63636 36363 63636 36363
63636 36363 63636 36363 63636 36363 63636 36363 63636 36363
63636 36363 63636 36363 63636 36363 63636 36363 63636 36363
63636 36363 63636 36363 63636 36363 63636 36222 94131 35807
27272 72727 27272 72727 27272 72727 27272 72727 27272 72727
27272 72727 27272 72727 27272 72727 27272 72727 27272 72727
27272 72727 27272 72727 27272 72727 27272 72727 27272 72727
27272 72727 27272 72727 27272 25031 65075 84119 99999 99999
99999 99999 99999 99999 99999 99999 99999 99999 99999 99999
99999 99999 99999 99999 99999 99999 99999 99999 99999 99999
99999 99999 99999 99999 99999 99999 99999 99999 99999 99999
99999 99999 99830 61240 54077 31759 99999 99999 99999 99999
99999 99999 99999 99999 99999 99999 99999 99999 99999 99999
99999 99999 99999 99999 99999 99999 99999 99999 99999 99999
99999 99999 99999 99999 99999 99999 99999 99999 99999 99999
37792 40588 59894 88859 99999 99999 99999 99999 99999 99999
99999 99999 99999 99999 99999 99999 99999 99999 99999 99999
99999 99999 99999 99999 99999 99999 99999 99999 99999 99999
99999 99999 99999 99999 99999 99999 99999 99765 68472 88372
27074 70599 99999 99999 99999 99999 99999 99999 99999 99999
99999 99999 99999 99999 99999 99999 99999 99999 99999 99999
99999 99999 99999 99999 99999 99999 99999 99999 99999 99999
99999 99999 99999 99999 99999 09976 07281 92626 42102 04519
99999 99999 99999 99999 99999 99999 99999 99999 99999 99999
99999 99999 99999 99999 99999 99999 99999 99999 99999 99999
99999 99999 99999 99999 99999 99999 99999 99999 99999 99999
99999 99999 99648 57932 42599 21622 89652 91716 36363 63636
36363 63636 36363 63636 36363 63636 36363 63636 36363 63636
36363 63636 36363 63636 36363 63636 36363 63636 36363 63636
36363 63636 36363 63636 36363 63636 36363 63636 36363 63634
97376 75910 84353 65491 94092 37458 18181 81818 18181 81818
18181 81818 18181 81818 18181 81818 18181 81818 18181 81818
18181 81818 18181 81818 18181 81818 18181 81818 18181 81818
18181 81818 18181 81818 18181 81818 18181 81262 44813 51260
22190 77609 77718 83803 63636 36363 63636 36363 63636 36363
63636 36363 63636 36363 63636 36363 63636 36363 63636 36363
63636 36363 63636 36363 63636 36363 63636 36363 63636 36363
63636 36363 63636 36363 63634 12085 39998 75472 67251 52257
48219 37636 36363 63636 36363 63636 36363 63636 36363 63636
36363 63636 36363 63636 36363 63636 36363 63636 36363 63636

36363 63636 36363 63636 36363 63636 36363 63636 36363 63636
36363 63636 35451 27249 80584 33192 27010 29252 84733 21381
81818 18181 81818 18181 81818 18181 81818 18181 81818 18181
81818 18181 81818 18181 81818 18181 81818 18181 81818 18181
81818 18181 81818 18181 81818 18181 81818 18181 81818 18178
08091 13338 08220 60020 00454 71333 95206 34663 63636 36363
63636 36363 63636 36363 63636 36363 63636 36363 63636 36363
63636 36363 63636 36363 63636 36363 63636 36363 63636 36363
63636 36363 63636 36363 63636 36363 63636 34823 44124 51939
26891 84500 91138 70707 75252 99399 99999 99999 99999 99999
99999 99999 99999 99999 99999 99999 99999 99999 99999 99999
99999 99999 99999 99999 99999 99999 99999 99999 99999 99999
99999 99999 99999 99999 99993 61845 82259 20170 42187 78210
95137 63245 46490 51399 99999 99999 99999 99999 99999 99999
99999 99999 99999 99999 99999 99999 99999 99999 99999 99999
99999 99999 99999 99999 99999 99999 99999 99999 99999 99999
99999 99999 97343 26339 82646 60423 00281 92032 91509 09033
02103 45999 99999 99999 99999 99999 99999 99999 99999 99999
99999 99999 99999 99999 99999 99999 99999 99999 99999 99999
99999 99999 99999 99999 99999 99999 99999 99999 99999 99988
89218 42681 44545 22857 47870 88961 79950 66706 09456 62599
99999 99999 99999 99999 99999 99999 99999 99999 99999 99999
99999 99999 99999 99999 99999 99999 99999 99999 99999 99999
99999 99999 99999 99999 99999 99999 99999 95337 89105 42583
81983 07896 06381 95680 99294 26072 26223 88399 99999 99999
99999 99999 99999 99999 99999 99999 99999 99999 99999 99999
99999 99999 99999 99999 99999 99999 99999 99999 99999 99999
99999 99999 99999 99999 99980 36404 47767 08257 93418 62042
15236 90003 66391 81162 81386 77927 27272 72727 27272 72727
27272 72727 27272 72727 27272 72727 27272 72727 27272 72727
27272 72727 27272 72727 27272 72727 27272 72727 27272 72727
27272 72727 18976 10932 22947 01014 42148 95581 04324 66449
27082 52737 06260 59563 63636 36363 63636 36363 63636 36363
63636 36363 63636 36363 63636 36363 63636 36363 63636 36363
63636 36363 63636 36363 63636 36363 63636 36363 63636 36328
47944 50714 05613 73042 92438 20648 89394 76060 33039 15512
09745 58272 72727 27272 72727 27272 72727 27272 72727 27272
72727 27272 72727 27272 72727 27272 72727 27272 72727 27272
72727 27272 72727 27272 72727 27272 72727 12335 25571 28479
08070 87568 62054 26392 96954 97604 08585 31870 87612 28007
27272 72727 27272 72727 27272 72727 27272 72727 27272 72727
27272 72727 27272 72727 27272 72727 27272 72727 27272 72727
27272 72727 27272 72727 27209 10513 42449 25002 57851 92601
19853 51864 31026 05399 00042 00415 74833 16356 36363 63636
36363 63636 36363 63636 36363 63636 36363 63636 36363 63636

36363 63636 36363 63636 36363 63636 36363 63636 36363 63636
36363 63636 09204 05235 81719 23828 01394 92038 91408 13025
05283 77500 56314 94650 06052 10632 72727 27272 72727 27272
72727 27272 72727 27272 72727 27272 72727 27272 72727 27272
72727 27272 72727 27272 72727 27272 72727 27272 72727 27156
54245 22924 68062 97829 42181 56049 48546 83619 84349 43893
50469 87346 34459 17359 99999 99999 99999 99999 99999 99999
99999 99999 99999 99999 99999 99999 99999 99999 99999 99999
99999 99999 99999 99999 99999 99999 99999 50200 78213 77716
55959 31714 24717 13624 28868 06712 22880 20560 80707 83278
04397 51199 99999 99999 99999 99999 99999 99999 99999 99999
99999 99999 99999 99999 99999 99999 99999 99999 99999 99999
99999 99999 99999 99999 99786 16215 84995 91490 68930 78097
73538 30269 55948 02231 04760 28810 55943 39592 08291 65279
99999 99999 99999 99999 99999 99999 99999 99999 99999 99999
99999 99999 99999 99999 99999 99999 99999 99999 99999 99999
99999 99999 08022 13617 52759 01188 74936 56619 47088 00461
48739 66089 95925 21290 30620 68413 98738 65919 99999 etc.

Isn't that a fine numerical dish for weekend pondering? This humongous number was calculated using the large-number software package VPCALC (http://pw1.netcom.com/~hjsmith/Calc/VPCalc.html) by Harry J. Smith.

Darrell Plank has since discovered a zebra irrational using a cube root:

$$\sqrt[3]{[(7^3)(10^{51}) + 7^5]/(11^3)}$$

or

$$[(343 \times 10^{51} + 16{,}807)/1331]^{(1/3)}$$

or

63636363636363636.
36363636363636363636363636363636
46
7575757575757575757575
7575757575757575757575
587
80808080808080808080808080808080808080808080
8542953423120089786756453423120089786756453423120075 . . .

Darrell writes to Dr. Oz:

In general, for the cube root, let your fraction be $[A^3 10^{3m} + d]/B^3$. Then the first three terms of the Taylor series are: $A/(B10^{-m}) + d/[(3A^2B)10^{2m}] - d^2/[9(A^5B)10^{5m}]$. (A Taylor series provides an approximation for the cube root.) Notice we are dividing by increasing powers of 10. These cause the "zebra" pattern to appear by shifting each fraction $3m$ digits to the right of the preceding fraction. An immediate implication is that using a larger value for m makes larger "bands" in the zebra irrational. In order to make this apparent, each fraction should be a repeating decimal with a fairly small period. The easiest way to do this is to ensure that the powers of d in the numerator divide out the powers of A in the denominator for the first few terms. Letting $A = 7$ and $d = 7^5$ in the example does this. B can be almost any small value (larger values can cause longer periods. We chose 11. In our example, the first fraction is $6.363636\ldots \times 10^{16}$. The second is $1.0393939\ldots \times 10^{-33}$, and the third is $1.69767676\ldots \times 10^{-83}$. Add these together and it possible to see how the zebra got his stripes.

Jason Earls has also explored zebra irrationals for higher exponents, such as with

$$\sqrt{(9/169 \times 100^{199} + (38 - 17 \times 199)/169}$$

which shows repeating motifs of clusters 230769, 410256, 213675, 296, 590693257359924026, and 914529. Another gem is

$$\sqrt{(9/64 \times 100^{155} + (92 - 22 \times 155)/64}$$

We call this zebra the "1481" zebra – not because this was the year the Russians finally defeated the Mongols – but for reasons that will become clear should any of you try to compute this gem of a number. As with all zebras, it is fascinating to watch the digits descend from order into apparent chaos. A most delicious pastime for anyone bored on a Saturday night is visually to search for the numerical-phase transition zone between order and inexorable madness. (I describe a similar class of numbers, Kevin Brown's "schizophrenic numbers," in my book *Wonders of Numbers*.)

Of course, the digits of all sorts of irrational numbers can exhibit obvious patterns. Consider, for example, Champernowne's number, which is irrational:

0.12345678910111213141516171819202021 . . .

(See how this is constructed by enumerating and concatenating the digits of successive integers?) The irrational number $0.10110111011110111110\ldots$ also exhibits patterns. In fact, the only patterns that irrational numbers cannot exhibit are ones that endlessly repeat themselves, like $0.272727272\ldots$. So, while it's true that many irrational numbers can exhibit "obvious patterns," we may still consider the "zebras" remarkable beasts worthy of study and periodic adoration.

Champernowne's number continues to fascinate me. It is "normal" in base 10, which means that any finite pattern of numbers occurs with the frequency expected for a completely random sequence. In fact, David Champernowne showed that not only will the digits 0 through 9 occur exactly with a 10% frequency in the limit, but each possible block of two digits will occur with 1% frequency in the limit, each block of three digits will occur with 0.1% frequency, and so on. Some cryptographers have noted that Champernowne's number does not trigger some of the traditional statistical indicators of nonrandomness. In other words, simple computer programs that attempt to find regularity in sequences may not "see" the regularity in Champernowne's number. This deficit reinforces the notion that statisticians and cryptographers must be very cautions when declaring a sequence to be random or patternless. It is not known if π and e are normal.

The continued-fraction representation of Champernowne's number is fascinating. As a reminder, a continued-fraction representation of a real number x is of the form:

$$x = a_0 + \cfrac{1}{a_1 + \cfrac{1}{a_2 + \cfrac{1}{a_3 + \ldots}}}$$

which can be represented in a compact notation as $x = [a_0; a_1, a_2, a_3, \ldots]$. Champernowne's number has a wild-looking continued fraction, with a sudden insertion of a huge numerator term:

[0; 8, 9, 1, 149083, 1, 1, 1, 4, 1, 1, 1, 3, 4, 1, 1, 1, 15,

457540111391031076483646628242956118599603939710457555000

066200439309026265925631493795320774712865631386412093755

03552094607183089984575801469863148833592141783010987,

6, 1, 1, 21, 1, 9, 1, 1, 2, 3, 1, 7, 2, 1, 83, 1, 156, 4, 58, 8, 54, . . .]

Another example of a normal number is the Copeland–Erdös constant:

0.23571113171923 . . . ,

which is obtained by concatenating the primes. In 1945, Arthur Copeland and Paul Erdös proved that this number is normal in all bases. Interestingly, this constant does not have huge terms in its continued fraction representation, which is in contrast to Champernowne's number.

Loki Jörgenson and Peter Borwein, of the Centre for Experimental and Constructive Mathematics at Simon Fraser University, use computer graphics when searching for patterns in numbers. For example, Figures F33.1 and F33.2 represent 1600 decimal digits of π and 22/7, respectively, both taken mod 2. In particular, the white cells □ represent the even digits, and the black cells ■ represent odd digits. The digits read from left to right, top to bottom, like words in a English text.

In Figure F33.1, as expected, the even and odd digits of π reveal no obvious patterns. The rationality of 22/7, an approximation for π often used by schoolchildren, is clearly apparent in Figure F33.2. The idea of visually representing apparent "randomness" in complicated data is not new, but slightly more sophisticated approaches might be used to hunt for patterns in the digit strings of irrational numbers or in the digits strings of continued fraction expansions, $[a_0; a_1, a_2, a_3, \ldots]$.

even
odd

F33.1. The first 1600 decimal digits of π **mod 2.** (Figure courtesy of Loki Jörgenson and Peter Borwein, Centre for Experimental and Constructive Mathematics, Simon Fraser University.)

even
odd

F33.2. The first 1600 decimal digits of 22/7 mod 2. (Figure courtesy of Loki Jörgenson and Peter Borwein, Centre for Experimental and Constructive Mathematics, Simon Fraser University.)

As an example of visualization applied to a slightly more complicated subject, consider that certain polynomials demonstrate amazing complexity when their zeros are appropriately plotted. Figures F33.3 and F33.4 display on the complex plane the complex zeros of all polynomials of the form:

$$P_n z = a_0 + a_1 z + a_2 z^2 + a_3 z^3 + \ldots + a_n z^n$$

for degree $n < 19$ and for $a_i = \{-1, +1\}$. Jörgenson and Borwein note that this image raises many questions, such as: Is the set fractal, and what is its boundary? Are there holes at infinite degree? How do the holes vary with the degree? What is the relationship between these zeros and those of polynomials with real coefficients in the neighborhood of $\{-1, +1\}$? Computer graphics allow us to magnify regions of these kinds of figures, and the graphics reveal an endless landscape of magnificent shapes and structures. Figure F33.4 shows the positive imaginary contributions shaded by their sensitivity to variation in the polynomial coefficients. (This looks particularly attractive in color.) In other words, the researchers examined the first derivative of the polynomial at each zero/root to see how large the derivative is. The shading indicates how quickly the value of the polynomial moves away from zero.

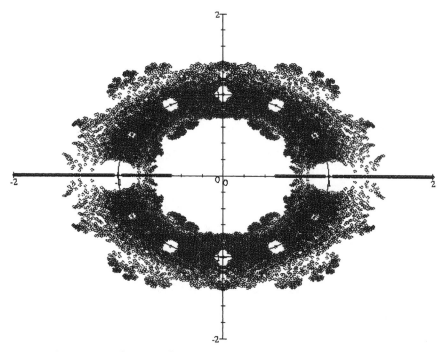

F33.3. Roots of polynomials of degree <19 for coefficients. This figure shows the complete distribution of zeros on the complex plane. (Figure courtesy of Loki Jörgenson and Peter Borwein, Centre for Experimental and Constructive Mathematics, Simon Fraser University.)

F33.4. "The Face of God Loki." Roots of polynomials of degree <19 for coefficients. With computer graphics and the judicious use of color, one can explore this amazing and infinite landscape at different magnifications. (Figure courtesy of Loki Jörgenson and Peter Borwein, Centre for Experimental and Constructive Mathematics, Simon Fraser University.)

For more information on this visualization work, see Loki Jörgenson, "Visible Structures in Number Theory," http://www.cecm.sfu.ca/~loki/. Also see Peter Borwein and Loki Jörgenson, "Visible Structures in Number Theory," *American Mathematical Monthly* 108(10): 897–911.

While on the subject of numbers with eye-catching patterns, Jason Earls enjoys experimenting with the *Rumpelstiltskin sequence,* which has a generating equation of the form $\beta \times (4n + 3)$. The equation yields numbers with all kinds of patterns:

n	y
1	7
2	121
3	1665
4	21109
5	255553
6	2999997
7	34444441
8	388888885
9	4333333329
10	47777777773
11	522222222217
12	5666666666661
13	61111111111105
14	655555555555549
15	6999999999999993
16	74444444444444437
17	788888888888888881
18	8333333333333333325
19	87777777777777777769
20	922222222222222222213
21	9666666666666666666657
22	10111111111111111111101

(Jason named the sequence the "Rumplestilskin sequence" after the German folktale in which Rumplestilskin spins straw into gold. This sequence similarly lets numerical experimenters turn ordinary-looking integers into numbers that delight the mind and eye.) The puzzle left for readers is to determine the correct expression to substitute for β so that the equation produces these numbers.

Here's a final digital delight. Consider the number 998. Determine its reciprocal:

$1/998 = 0.001002004008016032064128256513\ldots$

The result contains a sequence of powers of two: 1, 2, 4, 8, 16, ... What happens as the powers of two become larger and start overlapping? Notice that you can increase the spacing between the powers by adding 9's at the front of 998:

$1/99998 = 0.00001000020000400008000016\ldots$

To generate powers of 3, subtract 1 from 998:

$1/997 = 0.001003009027081243\ldots$

Can you generalize this approach to create other sequences of powers?

34. Creatures in Resin

Here is the largest block of repeats of which Dr. Oz is aware. (Can you prove that there is not a larger repeated block?) Try this problem on friends and offer them a reward if they can find the largest repeated block within five minutes.

Consider an *N*-by-*N* array of cells. The array is populated with *M* different symbols. What is the average size of the largest block of symbols you'd expect to see repeated in the array of cells? How would your answer change for cubical arrays and higher-dimensional arrays?

35. Prime-Poor Equations

Leonhard Euler (1707–83), the mathematician who discovered that $p = x^2 - x + 41$ produces many primes in a row, was a fascinating person. For one thing, he was the most prolific mathematician in history. Even when he was completely blind, he made great contributions to modern analytic geometry, trigonometry, calculus, and number theory. Euler published over eight thousand books and papers, almost all in Latin, on every aspect of pure and applied mathematics, physics, and astronomy. In analysis he studied infinite series and differential equations and introduced many new functions (e.g., the gamma function and elliptic integrals), and he created the calculus of variations. His notations, such as e and π, are still used today. In mechanics, he studied the motion of rigid bodies in three dimensions, the construction and control of ships, and celestial mechanics. Leonhard Euler was so prolific that his papers were still being published for the first time two centuries after his death. His collected works have been printed bit by bit since 1910 and will eventually occupy more than 75 large books.

Researchers have discovered a number of prime-poor equations for certain values of c that yield low densities $D(x_N)c$ for $x_N = 10,000$ (in other words, the density of primes is low as we scan from $x = 1$ to $x = 10,000$). Luckily for Dorothy, she was able to find an equation of the form $p = x^2 - x + c$ with the lowest density known for $x_N = 10,000$:

$$p = x^2 - x + 219,525$$

Its density $D(10,000)_{219,525}$ is 2.33%. Here are some example values for p using this formula:

x	p	x	p
1	$219,525 = 3 \times 5^2 \times 2927$	7	$219,567 = 3 \times 73,189$
2	$219,527 = 7 \times 11 \times 2851$	8	$219,581 = 23 \times 9547$
3	$219,531 = 3 \times 13^2 \times 433$	9	$219,597 = 3 \times 7 \times 10,457$
4	$219,537 = 3^3 \times 47 \times 173$	10	$219,615 = 3 \times 5 \times 11^4$
5	$219,545 = 5 \times 19 \times 2311$	11	$219,635 = 5 \times 13 \times 31 \times 109$
6	$219,555 = 3^2 \times 5 \times 7 \times 17 \times 41$	12	$219,657 = 3 \times 17 \times 59 \times 73$

Dr. Oz's friend Daniel Dockery notices something interesting about the values of x that produce prime numbers using this formula. In every case he tested for values of x, $0 \leq x \leq 5000$, the sum of the digits of x equals $2 + 3n$. This may also be true for larger values of x. For example, consider $x = 999,999,999,764$ (which produces the prime $999,999,999,527,000,000,275,457$ using the formula). This value of x has a digit-sum of 98, which is $2 + 3 \times 32$.

My friend Jason Earls asks if all primes of the form $n^2 - n + 219,525$ end with either 7 or 1 as their final digit. For example, here are some that end with 1: 220,931, 239,831, and 354,581. Will we ever see any that end in 3 or 9? If not, why?

Robert Sery, the world's expert on prime-poor equations, notes that for large enough values of c, much lower densities of primes should be attainable. This can be inferred from the prime-number theorem:

$$\lim_{n \to \infty} \frac{\pi(n)}{\left(\dfrac{n}{\ln n} \right)} = 1$$

where $\pi(n)$ is the actual number of primes less than n. In other words, the prime-number theorem states that the number of primes less than n is on the order of $n/(\ln n)$. For instance, there are 50,847,478 primes to $n = 10^9$, and the equation gives the approximate value 48,255,000. For large enough values of n, the densities of primes for prime-poor equations is expected to approach zero. According to Sery, "the object then of finding very low densities of primes becomes trivial if a reasonable limit is not set for the size of c."

On a related topic, create a triangle with 41 at the apex in the following manner (suggested to me by Ray Frame of Rockford, Illinois):

$$41$$
$$42\ \mathbf{43}\ 44$$
$$45\ 46\ \mathbf{47}\ 48\ 49$$
$$50\ 51\ 52\ \mathbf{53}\ 54\ 55\ 56$$
$$57\ 58\ 59\ 60\ \mathbf{61}\ 62\ 63\ 64\ 65$$
$$66\ 67\ 68\ 69\ 70\ \mathbf{71}\ 72\ 73\ 74\ 75\ 76$$
$$\cdots$$

Prime central

Why is it that the numbers in the central column yield so many primes? Does the triangle geometrically recast Euler's formula $p = x^2 - x + 41$? Why should such a triangle yield so many primes? Do you think Euler knew of this triangle? The central column works for values up to the column value of 1681 which, interestingly enough, is 41^2.

The following BASIC code is a starting point for those of you would like to generate prime numbers.

```
10  REM Generate Prime Numbers
11  REM Since 2 is the only even prime, check only
12  REM odd numbers.  Divide each odd number by all primes
13  REM that are found.
14  DIM A[600]
20  Print "Here is a list of prime numbers:"
22  R=1
25  A[1]=2
30  P=1
35  FOR X=3 TO 600 STEP 2
40      FOR Y=1 TO R
41          REM Is number divisible by previous primes?
45          IF INT(X/A[Y])*A[Y]=X THEN 100
50      NEXT Y
55      R=R+1
60      A[R]=X
65      IF P > 6 THEN 85
70      P=P+1
75      PRINT X;
80      GOTO 100
85      P=1
90      PRINT X
100 NEXT X
110 END
```

For more information on prime-poor equations, see Robert S. Sery, "Prime-poor equations of the form $i = x^2 - x + c$, c odd," *Journal of Recreational Mathematics* 30(1) (1999/2000): 36–40.

36. Number Satellite

Figure F36.1 shows one solution. Are there others? If you are a teacher, have your students create similar "satellite" puzzles to test one another. For similar kinds of puzzles in color, see my *Mind-Bending Puzzle* calendars, published by Pomegranate.

B.C. MANSFIELD

F36.1. Number satellite. (Illustration by Brian Mansfield.)

37. Flatworm Math

The algorithm for flatworm numbers looks like this:

```
input x
  repeat
    x ← trunc(2x)
    output x
  until x= a previous x
```

The recipe is simple. Using a computer, truncation of the last two digits is obtained by applying the mod function at 100, that is, taking the remainder of x on division by 100. The algorithm ends when the variable x repeats a value taken earlier. (The number of steps needed to return to the initial number is called the "path length.") After the sequence repeats a number, the sequence of values will cycle endlessly and there is no point in generating more values. Can you think of an efficient way that a computer program might detect the repetition of x-values?

Odd starting numbers will never return to themselves, so we test only even numbers. You will find, after much exploration that there are, remarkably, only three possible outcomes if you start with any even number: 1) the path never returns to the starting point, but rather executes a repeating loop; 2) the path length is 20; or 3) the path length is 4. In particular note that if the starting number is a multiple of 20, it will take 4 iterations to repeat; if the starting number is a multiple of 4, it will take 20 iterations; otherwise, it will loop indefinitely.

If you like, you can begin to analyze Flatworm Math by noting that the truncation to two digits implies addition modulo 100. Denote the starting integer as y. It is easy to see that after n successive doublings we have $(2^n)y$ mod 100. For the sequence to repeat, we require this quantity to equal y for some n. Therefore, to produce a return cycle requires $y[2^n - 1]$ mod $100 = 0$. (Note that Flatworm Math has similarities with a class of problems used in the field of pseudorandom number generators based on multiplicative–congruential techniques of Donald Knuth.)

Have fun and explore! Do flatworm sequences always return to the starting point for related problems, for example, where 3 is used as the multiplicative factor instead of 2? What graphics can you develop to show patterns? One beautiful way to summarize the behavior for all cases is to draw an iteration diagram. You can place each of the starting numbers on the x-axis and plot their trajectories on the y-axis. Place each of the numbers between 0 and 99 on a sheet of paper in the form of a dot. Now apply the iteration (repetition) rule you wish to investigate and draw an arrow from the number to which the rule was applied to the number that resulted. Soon enough the page will fill with a confusion of arrows, but a little disentangling and redrawing will produce an elegant figure. Here are all possible flatworm sequences simultaneously visible!

As far as programming projects are concerned, the holy grail would certainly have been attained by the reader who found a novel way to automate the graph process. Imagine merely feeding in two parameters that specified the rule, then watching the screen blossom with the diagram. Here are some references for you to consider:

- Knuth, D. (1981). *The Art of Computer Programming*, vol. 2. 2d ed. Reading, Mass.: Addison–Wesley.
- Pickover, C. (1998). "Wormy algebra," *Odyssey*. 7(6) (September): 37.
- Pickover, C. (1992). *Computers and the Imagination*. New York: St. Martin's Press.
- Pickover, C., and Runger, G. (1995). "The 2N problem," *Journal of Recreational Mathematics* 27(3): 172–4.
- Pratt, L. (1992). "A note on earthworm algebra and computer graphics," *Computers & Graphics*. 16(3): 339–40.

38. Regolith Paradox

The answer is (a). Each triplet must repeat exactly two symbols, regardless of size. The paradox is: Why can't anyone solve the puzzle when it seems so obvious in hindsight?

39. 🐾 🐎 🐎 🐾 🐾 🐾 🐎 🐎 🐎 🐎

Dr. Akhlesh Lakhtakia has studied the interesting generating formula for u_n:

$$u_n = 2n - \{(1 + \sqrt{8n - 7})/2\}$$

where $\{\varepsilon\}$ denotes the greatest integer not exceeding ε. As n approaches infinity, u_n/n becomes almost $2 - [1/n - 0.5\sqrt{(8/n)} - (7/n)(1/n)]$. Because (a/n) approaches 0 as n approaches infinity, the ratio u_n/n is 2 for large n. For example, for $n = 1000$ we have $u_n = 1955$, and the ratio is 1.955. Dr. Oz encourages you to plot a graph of u_n vs. n and experiment with other Connell-like sequences such as those produced by taking two odd numbers, followed by four evens, followed by six odds, and so on: 🐾 🐾 🐎 🐎 🐎 🐎 🐾 🐾 🐾 🐾 🐾 🐾.

Perhaps a better way of demonstrating the limiting behavior discussed in the previous paragraph is as follows. Dr. Oz makes use of the fact that the Connell sequence is composed of subsequences:

subsequence number	subsequence
1	1
2	2, 4
3	5, 7, 9
4	10, 12, 14, 16

Note that there are q members in subsequence q. Note also that the last element of each subsequence is q^2; thus the value of the last Connell element in any subsequence can be expressed by $u_{(1/2)q(q+1)} = q^2$. For example, if $q = 2$, we have $u_3 = 2^2$. Dr. Oz also notes that $u_{(1/2)q(q+1)-p} = q^2 - 2p$ $\{q = 1, 2, 3, \ldots, p = 0, 1, 2, \ldots, q - 1\}$, where p can achieve q values. Let us consider the ratio:

$$[u_{(1/2)q(q+1)-p}]/[\tfrac{1}{2}q(q+1) - p] = [q^2 - 2p]/[\tfrac{1}{2}(q^2 + q - 2p)]$$
$$= 2 \times \{1/[1 + q/(q^2 - 2p)]\}$$

Since $0 \le p \le q - 1$, the limit of this ratio as q approaches infinity is 2. Note also for q, the sum of the first $\tfrac{1}{2}q(q+1)$ Connell numbers divided by the sum of the first $\tfrac{1}{2}q(q+1)$ integers also tends to 2. To be more precise, note that

$$\sum_{n=1}^{n=(1/2)q(q+1)} u_n = \tfrac{1}{12}q(q+1)(3q^2 - q + 4)$$

Let $S_q = 1 + 2 + 3 + \ldots + \tfrac{1}{2}q(q+1)$, that is, S_q is the sum of all integers from 1 to $\tfrac{1}{2}q(q+1)$. Then

$$S_q = \tfrac{1}{2}[\tfrac{1}{2}q(q+1)][\tfrac{1}{2}q(q+1)+1]$$
$$= \tfrac{1}{8}q(q+1)q^2+q+2)$$

Therefore,

$$\frac{1}{S_q}\sum_{n=1}^{n=(1/2)q(q+1)} u_n = \tfrac{2}{3}(3q^2-q+4)/(q^2+q+2)$$
$$= \tfrac{2}{3}\{3-[2(2q+1)]/[q^2+q+2]\}$$

In the limit, as q approaches infinity, this ratio also goes to 2. Further reading:

- Connell, I. (1959). "Elementary problem E1382," *American Mathematical Monthly* 66(8)(October): 724.
- Connell, I. (1960). "An unusual sequence," *American Mathematical Monthly* 67: 380.
- Lakhtakia, A., and C. Pickover (1993). "The Connell sequence," *Journal of Recreational Mathematics* 25(2): 90–2.

40. Entropy

If these experiments are done with large spheres in a sea of smaller ones, researchers find that the large particles eventually find each other and stick together. By staying close together or by moving into a corner, the large particles free up a small amount of space for the smaller particles to move around in, which increases the overall disorder of the system. The smaller particles become more disorderly. Imagine a scene in which people who don't want to dance crowd against the walls to make more room on the dance floor for the wild dancers. Because each particle contributes equally to the overall entropy, the increase in entropy from the many small spheres more than compensates for the reduction in entropy from the few larger spheres. This has been experimentally verified for micrometer-sized spheres moving in water. For example, spheres floating around in a solution are forced into a crystalline order along a glass wall when smaller spheres are added. The big spheres hug as much wall as possible to make more room for the little ones. Therefore, Dorothy should expect the three large Mr. Plex's to be clumped together against a wall.

Usually, people think of disorder as always increasing in a random system. For example, if you mix 100 blue marbles and 100 red marbles in a bag and shake it for a long time, the marbles become randomly mixed. But sometimes entropy can be a force for organization. Under certain circumstances, the marbles will, in effect, segregate. An increase in entropy in one part of a system can force another into greater order. As discussed in David Kestenbaum's *Science* magazine article "Gentle force of entropy bridges disciplines" (*Science* 279 [1998]: 1849), this ordering effect of entropy has staggering implications in many fields of science.

41. Animal Gap

The answer is (a). The pattern 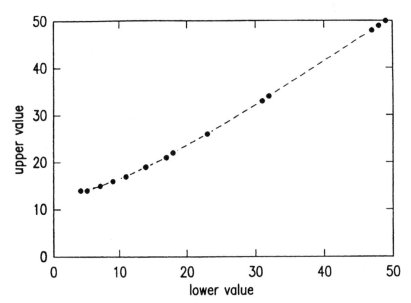 repeats, starting at the upper left and proceeding downward. When the bottom of a column is reached, continue on with the pattern along the column to the right.

Did you notice that we obtain the same answer if the repeating sequence is 🦇 🦆 🦆 🦆 🐦 🐦 🦇 🦇 and we proceed left to right along rows? Is this an accident, or will it happen any time we write a repeating sequence of eight symbols in an N-by-N array? Search for other patterns.

42. Arranging Alien Heads

As a final observation, note that because $\sum_{n=1}^{t} n = [t(t+1)]/2$, we can express consecutive integer partitions in terms of the lower l and upper u value in the summation: $\frac{1}{2}u(u+1) - \frac{1}{2}l(l+1) + 1 = a$. This means that values of l and u must lie on a hyperbolic curve of the form $u^2 - l^2 + u + l = 2a$. For example, for $a = 21 = 11 + 10$ we obtain $121 - 100 + 11 + 10 = 42$. Figure F42.1 shows a hyperbolic distribution for all lower and upper bounds for nearby a values ($95 \le a \le 100$). When placed in a standard form, it becomes easier to see that $u^2 - l^2 + u + l = 2a$ represents a hyperbola shifted in the x and y axes: $(u + \frac{1}{2})^2/(2a) - (l - \frac{1}{2})^2/(2a) = 1$, with a distance from its center to focus equal to $\sqrt{4a}$ and with an asymptote slope of 1.

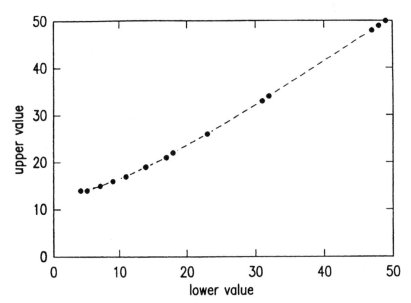

F42.1. Hyperbolic distribution for all lower and upper bounds for nearby a values ($95 \le a \le 100$).

Dr. Oz urges you to experiment with similar graphs for consecutive multiplied integers. For example, if $a = 30$, then one such product is $5 \times 6 = 30$. To encourage your involvement, Dr. Oz gives pseudocode here for constructing the graph in Figure 42.1 and for partitioning a into consecutive integers. You can convert this rough outline into the computer language of your choice. The following computer code permits further exploration.

ALGORITHM: Compute Partition Graph for Consecutive Integer Sums
This pseudocode can be converted to the language of your choice.

Input:

```
        start – the smallest alpha value shown on the x-axis
        stop  – the largest alpha value shown on the x-axis

scale = 100/stop;      /* this scales the x-axis for the plot */
DO a = start to stop; /* scan a range of a values        */
   looptop=a/2+1;      /* need to scan only to half of a */
   DO i = 1 to looptop; /* search for consec. partitions for   */
     sum=0; top=i;       /*   a particular alpha value        */
     again: sum = sum + top;
     top = top + 1;
     if sum < a then goto again;
     if sum = a then do;
       /*print out lower and upper values*/
       PrintNumbers(a,i,top-1);
       DO k = i to top-1;
          PlotDotAt(a*scale,k*scale); /* plot a dot at x,y */
       END; /* k */
   END;   /* i */
END; /*start to end*/
```

43. Ramanujan Congruences and the Quest for Transcendence

For decades after Ramanujan's findings, researchers hunted in vain for additional congruences like the ones in this chapter, and most of their searches met with failure. Then suddenly, in 2000, mathematician Ken Ono of the University of Illinois proved there are infinitely many such relationships! One of the relationships Ono discovered was that congruences exist for all primes larger than 5 – although he provided only one explicit example. In the next congruence the starting number is 111,247, with each successive step to the next integer going up by $59^4 \times 13$. The corresponding partitions are all multiples of the prime number 13.

An undergraduate student, Rhiannon Weaver, formerly at Pennsylvania State University, subsequently found an algorithm to obtain more than 70,000 new congruences! The resulting numbers are often monstrously large. Recall, our previous congruences looked something like $p(11N+6) \equiv 0 \pmod{11}$ for $N \geq 0$. Rhiannon Weaver discovered:

$$p(11{,}864{,}749N + 56{,}062) \equiv 0 \pmod{13} \quad \text{for } N \geq 0$$

$$p(14{,}375N + 3{,}474) \equiv 0 \pmod{23} \quad \text{for } N \geq 0$$

Despite all the flurry of research, no one knows whether there are congruences that involve multiples of 2 or 3! Ono says that partitions are "much more than just counts of how to add up numbers. They are a vehicle for testing some of the most important conjectures about mathematical objects that we can barely get a handle on otherwise." In some way, what prime numbers are to multiplication, partitions are to addition.

Some more partition trivia: 666 has 11,956,824,258,286,445,517,629,485 partitions. Note that $p(666) = 11{,}956{,}824{,}258{,}286{,}445{,}517{,}629{,}485 = 5 \times 11 \times 709 \times 306{,}624{,}548{,}231{,}476{,}997{,}503$. Also for partition fans,

$$p(1000) = 24{,}061{,}467{,}864{,}032{,}622{,}473{,}692{,}149{,}727{,}991$$

For more information on Ramanujan congruences see:

- Danahy, Anne (2000). "Undergraduate creates concrete proof from complex mathematical theory," Penn State Intercom Online, University Relations, http://www.psu.edu/ur/archives/intercom_2000/May18/math.html.
- Peterson, Ivars (2000). "The power of partitions," *Science News* 157(25) (June 17): 396–7 (see http://www.sciencenews.org/20000617/bob2.asp for a reprint).

Srinivasa Ramanujan (1887–1920), who started working as a clerk in a Madras post office accounting department, became India's greatest mathematical genius and one of the greatest twentieth-century mathematicians. Ramanujan made substantial contributions to the analytical theory of numbers and worked on elliptic functions, continued fractions, and infinite series. He came from a poor family, and his mother took in boarders, which created a crowded home. Ramanujan was very shy and found it hard to speak. He excelled in math but usually failed all his other courses. When he was thirteen, he borrowed a high-school student's math book and mastered it in a week. Because he was deprived of manuals that could teach him about rigorous proofs, Ramanujan developed rather strange methods to establish mathematical truths. Mathematician G. H. Hardy of Trinity College remarked:

His ideas as to what constituted a mathematical proof were of the most shadowy description. All his results, new or old, right or wrong, had been arrived at by a process of mingled argument, intuition, and induction, of which he was entirely unable to give any coherent account.

Ramanujan, although self-taught in mathematics, was given a fellowship to the University of Madras in 1903, but the following year he lost it because he devoted all his time to mathematics and neglected his other subjects. Professor Hardy invited him to Cambridge on the basis of Ramanujan's historic letter that contained some hundred theorems. Hardy, a leading expert in analysis, found himself dealing with a collection of formulas completely unfamiliar to him:

These relations defeated me completely; I had never seen anything in the least like them before. A single look at them is enough to show that they could only be written down by a mathematician of the highest class.

Later in life Ramanujan was quite sick due to malnutrition and tuberculosis. However, neither physicians nor his family could persuade him to stop his studies. He returned to India in February 1919 but died in April 1920 at the age of 32. During that period he wrote down about six hundred theorems on loose sheets of paper, which were discovered only in 1976 by Professor George Andrews of Pennsylvania State University, who termed these the "Lost" Notebook of Ramanujan. Ramanujan's formulas often took central places in modern theories of algebraic number theory, and today scholars wonder how he could envision such equations when he didn't have any of the supporting knowledge to understand them.

One biographer said about Ramanujan: "Like Albert Einstein, who toiled as a clerk in a Swiss patent office while evolving his Special Theory of Relativity at odd hours, Ramanujan worked as a clerk at a port authority in Madras, spending every spare moment contemplating the mathematical face of God."

For more on Ramanujan's life, see:

- Berndt, B., and R. Rankin (1995). *Ramanujan: Letters and Commentary*. Providence, R.I.: American Mathematical Society.
- Gindikir, S. (1998). "Ramanujan the phenomenon," *Quantum* 8 (March–April): 4–9.
- Kanigel, R. (1991). *The Man Who Knew Infinity: A Life of the Genius Ramanujan*. New York: Charles Scribner's Sons.

🐜 🐜 🐜

44.   Getting Noticed

After polling top scientists around the country, I found several amazing answers. Of course, there is no "right" answer." Before looking at the solutions, what would you do if you were trapped in an ant's body?

Many scientists suggest that sound would be one way to communicate. Some ant species use sounds to communicate alarm. Carpenter ants drum their heads on the floor of their chambers, and leafcutter ants and harvester ants make squeaking sounds if their nest caves in. Nestmates follow these sounds to find and rescue the trapped ants.

If you could somehow make sounds, even by primitive methods like tapping a foot, and if you could convince someone to place a high-powered microphone near you, it might be possible to communicate by Morse code or other means. The first step would be to spell out a simple message, such as "microphone" to get people to understand your needs. If dye or ink was available, perhaps you could spell a message, but many dyes and inks could be toxic. Perhaps if you could tear pieces of paper with your strong jaws, the tiny paper fragments could be arranged to spell out a message. It may be possible to indicate certain mathematical relationships, but how many people in the world today, selected at random, would recognize prime numbers, the Pythagorean theorem, or binary numbers?

Even if people could recognize that you are a human trapped in an ant's body, could you live a happy life? Imagine how alien the world would look to you through your new senses. Imagine how scared you would be, vulnerable to other insects and being stepped on. Is there anything you could do to avoid sinking into a state of incredible depression or despair? Is there some way you could keep your spirits from slipping into the abyss? Is there some way you could enjoy the strange new world around you, by meditating, or by ingesting some hallucinogenic ergot? If you believe in God, would your soul still be crammed "within" the prison of your segmented body?

Josh, a twelve-year-old neighbor down the street, wrote to me:

I know that some types of ants conduct organized warfare. That might catch attention of a child or someone walking by. Then I would find a way to spell out HELP, and I also know that most ants can carry many times their weight. I would get small berries or pebbles around the area and do it as quickly as I could to get people's attention. This would give me more time to spell out words and finally tell them that I am really a human. They may not believe it at first, or at all for that matter, but they might eventually be so amazed that they have to believe that I am really a human and finally try to cure me.

Of course, it would seem impossible for an ant's minuscule brain to hold one's memories and exhibit one's cognitive abilities. What other practical or scientific difficulties do you find in this scenario? Because your visual system (e.g., lens and retina) is smaller, would this affect your sense of sight?

When I think about the possibility of inhabiting another animal's body or traveling to alien worlds in outer space, I always remind myself that the most exotic journey would not be to see a thousand different worlds, but to see a single world through the eyes of a thousand different aliens. Not only do I mean this in the symbolic sense of viewing the world from various alien perspectives, but I also literally mean seeing through eyes sensitive to strange, nonvisible parts of the electromagnetic spectrum, seeing in all directions simultaneously, or seeing events that are so quick that they are a mere blur to the human eye. By studying the creatures of Earth, we can hypothesize on the diversity of alien eyes and visual perceptions. Aliens would no doubt see a different world than we do. To best understand this, consider the Indian luna moth, which has a wingspread of about 10 cm (4 in.). To our eyes, the male and female moths are both light green and indistinguishable from each other. But the luna moths themselves perceive in the ultraviolet range of light, and to them the female looks quite different from the male. Other creatures have a hard time seeing the moths when they rest on green leaves, but luna moths are not camouflaged to one another since they see each other as brilliantly colored.

On Earth, bees can also see into the ultraviolet, although they do not see as far as we do into the red range of the spectrum. When we look at a violet flower, we do not see the same thing that bees see. In fact, many flowers have beautiful patterns that only bees can see to guide them to the flower. These attractive and intricate patterns are totally hidden to human perception.

Bees also see the world differently from us because of their remarkable flicker-fusion rate. "Flicker-fusion" refers to the number of frames per second at which sequential images are no longer seen as separate. Humans can distinguish from 16 to 24 flickers per second. Films, viewed as motion pictures, usually show 24 still frames per second. If an alien had the flicker-fusion of a bee, the alien could see 265 separate flickers per second before fusion takes place. Our movies would seem like a slide lecture to them. As a result of their higher flicker-fusion, bees can see objects moving at higher speeds than we can see because to us the image becomes a blur or eventually invisible.

Imagine aliens whose gestures are so quick that we can't see them but that the aliens would have no trouble seeing. Imagine what it would be like for us to see the vibration of fly's wing in flight or the intricate array of droplets splashed when a raindrop hits a puddle.

If we were able to extend our current senses in range and intensity, we could glimpse alien sense-domains. If we had grown up from birth with these visual skills, our species would be transformed into something quite unusual. Our art would change, our perception of human beauty would change, our ability to diagnose diseases would change, and even our religions would change. If only a handful of people had these abilities, would they be hailed as messiahs?

45. Juggler Numbers

Juggler numbers can reach amazing heights before returning to 1. Here are the first few terms of a Juggler behemoth starting with 193, which takes 74 steps before returning to 1.

1 193 (Starting number)
2 2681
3 138817
4 51720650
5 7191
6 609795
7 476185085
8 10391151638843
9 33496198677403032405
10 193862266440176814000739774971853571544597469727317610662626120110666679467
11 788605665777872333941479874994035776759769384273765832159588845286
12 8880347210429738224951866074446955
13 264633560309716104220858647037719228740400480880677
14 136134242643398481765860616637779751344646303632658099373147841507505327119

Dr. Oz performed the computation on an IBM 3090 using special-purpose software that allows high-precision exponentiation. To compute multidigit square roots for a number x, Dr. Oz iterated the following computer command several hundred times: answer = 0.5 * (x/answer + answer).

Cornelius Groenewoud of Bartow, Florida, wrote Dr. Oz the following letter:

Dear Sir: You have essentially defined the Juggler sequences by

if x is even   then $x \leftarrow [x^f]$
                     else $x \leftarrow [x^g]$

until $x = 1$

where $f = 0.5$ and $g = 1.50$. I think proving that every such sequence ends in 1 will be very difficult. The termination is very sensitive to the choice of f and g. As an example, let's always choose 5 as the initial value of x and then use different values of f and g very near to your choices of $f = 0.50$ and $g = 1.50$. Note what happens:

f	g	Members of sequence
0.55	1.45	5, 10, 3, 4, 2, 1
0.54	1.46	5, 10, 3, 4, 2, 1
0.53	1.47	5, 10, 3, 5, repeats
0.52	1.48	5, 10, 3, 5, repeats
0.511	1.489	5, 10, 3, 5, repeats
0.510	1.490	5, 11, 35, 199, 2662, . . .
		a total of 18 steps ending in . . . 4, 2, 1
0.50	1.50	5, 11, 36, 6, 2, 1 repeats
0.49	1.51	5, 11, 37, 233, 3755, 249,839, 141,405,711, etc.
0.48	1.52	5, 11, 38, 5 repeats
0.473	1.527	5, 11, 38, 5 repeats
0.472	1.528	5, 11, 39, 269, 5160, 56, 6, 2, 1
0.471	1.529	5, 11, 39, 270, 13, 50, 6, 2, 1
0.47	1.53	5, 11, 39, 271, 5277, 495,738, 475, 12,454, 84, 8, 2, 1
0.46	1.54	5, 11, 40, 5, repeats
0.45	1.55	5, 12, 3, 5, repeats

You may wish to make a 3-D plot showing the relationship among f, g, and the number of steps before returning to 1.

Others have written to Dr. Oz regarding the Juggler sequence. James Beauchamp of Quebec, Canada, notes that, in some versions of BASIC and on some computers, the statement x = INT(36^(1/2)) gives the result 5, which is false. He suggests SQR be used instead of exponentiation to the 1/2 or 0.5 power.

On June 27, 1992, Harry J. Smith computed the world-record holder for Juggler numbers. Printing the entire number would require the same number of pages as in this book. The starting number is 48,443. The sequence starts with $J(0) = 48,443$, $J(1) = 10,662,183$, $J(2) = 34,815,273,349$, and it reaches its peak at $J(60)$, which has 972,463 digits. Alas, as hypothesized for all Juggler numbers, Smith's mighty sequence meets its demise, at $J(157)$, at which point the Juggler sequence has decayed back to 1. Harry Smith writes:

The research and development required to compute this number on a PC was quite interesting. I knew from my earlier work that $J(0) = 48,443$ produced a very large number because my program ran out of memory while computing this sequence. The numbers were getting too large to be held in a Turbo Pascal array. In Turbo Pascal, arrays are limited to 65,536 bytes.

Harry Smith finally wrote a program using the object-oriented programming language Turbo C++ for Windows Version 3.0, by Borland International, in order to break the world record and to perform the necessary multiple precision-integer arithmetic required. His computer was a mere IBM AT–compatible 33MHz 486 with 16 megabytes of RAM and a 330-megabyte hard disk drive. To overcome his computer's limitations, Smith used Fast Fourier Transform techniques to speed up the large-number arithmetic, and a variant of Newton's method to compute the square roots quickly. As a result, the computation of the world-record Juggler number required 28 hours. Smith is happy to answer questions regarding his software and methodology. Contact him at: Harry J. Smith, 19628 Via Monte Dr., Saratoga, CA 95070.

Roger Caws from West Sussex, United Kingdom, has proposed and studied "reverse" Juggler sequences. He writes, "It is possible to map all possible reverse Juggler sequences by making $j(1) = 1$ and then producing all possible next values considering the following: $j(n + 1)$ is the set of odd integers from $j(n)^{2/3} \ldots < (j(n) + 1)^{2/3}$ + set of even integers from $j(n)^2 \ldots < (j(n) + 1)^2$.

Juggernaut is an informal newsletter published by the Juggler Geometry Club. It consists of a large collection of letters written by students and researchers discussing both practical and theoretical aspects of juggler geometry. Inquire at: Dr. Oz, c/o Cliff Pickover, P.O. Box 549, Millwood, New York 10546-0549 USA.

For those of you who want to use a BASIC program to explore smaller Juggler numbers, this following code should provide a useful starting point:

```
10 REM Compute Juggler Numbers
15 REM For extremely large numbers,
16 REM other methods may have to be used.
20 INPUT "Enter starting number (e.g. 77) >"; N&
30 MAXVAL& = 0
40 PATH% = 0
50 WHILE N& > 1
60    PATH% = PATH% + 1
70    IF N& > MAXVAL& THEN MAXVAL& = N&
80    IF N& MOD 2 = 0 THEN N& = INT(SQR(N&)) ELSE N& = INT(N&^1.5)
85    REM = INT(SQR(N&*N&*N&))
90    PRINT N&,
100 WEND
110 PRINT
120 PRINT "Path length:"; PATH%; "Maxium Value:"; MAXVAL&
130 END
```

There is a vast literature on the related $3n + 1$ problem. Here are some favorite references:

- Crandall, R. (1978). "On the '$3x + 1$' problem," *Mathematics of Computation* 32: 1281–92.
- Dodge, C. (1969). *Numbers and Mathematics*. Boston: Prindle, Weber, & Schmidt.
- Garner, L. (1981). "On the Collatz $3n + 1$ problem," *Proceedings of the American Mathematical Society* 82: 19-22.
- Hayes, B. (1984). "Computer recreations: On the ups and downs of Hailstone numbers," *Scientific American* 250: 10–16.
- Lagarias, J. (1985). "The $3x + 1$ problem and its generalizations," *American Mathematical Monthly* 92(1) (January): 3–23.
- Lagarias, J., and A. Weiss (1992). "The $3x + 1$ problem: two stochastic models," *The Annals of Applied Probability* 2(1): 229–61.
- Leavens, G., and M. Vermeulen (1992). "$3x + 1$ search programs," *Computers and Mathematics with Applications* 24(11): 79–99.
- Pickover, C. (1989). "Hailstone ($3n + 1$) number graphs," *Journal of Recreational Mathematics* 21(2): 112–15.
- Silva, T. (1999). "Maximum excursion and stopping time record-holders for the $3x + 1$ problem: computational results," *Mathematics of Computation* 68(225): 371–84.
- Wagon, S. (1985). "The Collatz problem," *Mathematical Intelligencer* 7: 72–6.
- Wirsching, G. (1999). *The Dynamical System Generated by the 3N+1 Function* (Lecture Notes in Mathematics). New York: Springer–Verlag.

Eric Roosendaal is conducting a massive collaborative computer search of the $3n + 1$ numbers at http://personal.computrain.nl/eric/wondrous/ in order to determine if all starting numbers eventually "fall" down to 1. In August 1999 the first major international conference on the $3n + 1$ problem took place at Katholische Universität Eichstätt in Eichstätt, Germany.

46. Friends from Mars

Dr. Oz conducted a study of almost five hundred people with regard to this column-connection problem, and he asked individuals to time themselves as they attempted to arrive at a solution. About 20% said this problem was impossible to solve. Those that could solve it usually did so in under two minutes, and there was little correlation between a person's ability to solve the puzzle and age (ages ranged from 20 to 60). The problem is in fact solvable, and the solution is left as an exercise for you. If you cannot solve the problem, don't think about it for a day, and then return to it. Many people find it easier to solve this on their second attempt a day later. A computer could probably solve this class of problems faster than a human; however, humans have one advantage in that they have the ability to discard bad attempts rather quickly. Write a computer program to place aliens randomly to create new and unusual "wiring" problems, or you can create new puzzles like this with pencil and paper.

Psychologists have long been interested in the relationship between visualization and the mechanisms of human reasoning. Is it significant that people find the puzzle easier to solve after returning to it a day later? Is there any correlation in a person's ability to solve the puzzle with gender, profession, IQ, musical ability, or artistic ability?

This type of problem raises questions that pertain to the mathematical field of graph theory – the study of ways in which points can be connected. Graphs often play important roles in circuit design.

47. Phi in Four 4's

Phil Hanna from New York had some excellent approximations, including:

$$\frac{4}{\sqrt{4} + \sqrt{4}/4} = 1.6$$

For Contest 2, he found:

$$\sqrt{\sqrt{4 + \sqrt{\sqrt{4 \times 4 \times 4}}}} \sim 1.61651660$$

which differs from ϕ by 1.517×10^{-3}.

Leopold Travis from Brandeis University derived the following unusual formula. Let $s(x)$ denote the square root of x, and let

$$l = s(s(s(44))) \times s(s(s(s(s(s(s(s(s(s(s(s(s(4^{**}(4!))))))))))))))$$

where ** denotes exponentiation. Then $l = 1.61792833086266\ldots$ and $\phi - l = 0.00010565788722\ldots$.

Ken Shirriff, from California, noted that for Contest 1, one can obtain results arbitrarily close to ϕ because infinitely repeated square roots yield numbers approaching 1:

$$\ldots \sqrt{\sqrt{\sqrt{\sqrt{4}}}} \ldots \to 1$$

Therefore,

$$\phi = \frac{1 + \sqrt{4 + 1}}{\sqrt{4}} \quad \text{exactly,}$$

where 1 can be approximated as closely as desired with a single 4 and many square roots.

For Contest 2, the best Ken could do was:

$$1.618644 = \frac{4}{(.4 \times 4!)^{.4}}$$

with an error of 0.000611.

Similarly, the following, from Paul Leyland of the UK can be used to compute ϕ to arbitrarily close precision:

$$\phi = \frac{\sqrt{\sqrt{4/.4}} + \ldots \sqrt{\sqrt{\sqrt{4}}} \ldots}{\sqrt{4}}$$

To see this, note that

$$\sqrt{\frac{\sqrt{4}}{.4}} = \sqrt{\frac{2}{2/5}} = \sqrt{5}$$

And the limit of $4^{1/2^n}$ is 1 as n goes to infinity.

David G. Caraballo of Princeton discovered $(0.4 + 4^{-4}) \times 4$. This equals $1.615625\ldots$, and therefore differs from the golden ratio by around 0.00240.

The ultimate winner of the contest, however, is Brian Boutel, of the Victoria University of Wellington, New Zealand, who was the first person to find an *exact* solution:

$$\phi = \frac{\sqrt{4} + \sqrt{4! - 4}}{4}$$

because ϕ, as noted previously, is defined as $(1 + \sqrt{5})/2$.

When Dr. Oz extended the contest to computing ϕ using five 5's, six 6's, and so on, David G. Caraballo from Princeton, New Jersey, noticed that exact solutions can be computed using five 5's and seven 6's:

$$\frac{5 + 5 \times \sqrt{5}}{5 + 5} \quad \text{and} \quad \frac{6 + 6 \times \sqrt{6} - 6/6}{6 + 6}$$

He proposes a general solution involving the integer k. We need at most $(2k - 5)$ k's to give an exact value using only k's and the operations previously described. Can you prove this?

Other very imaginative answers for ϕ in five 5's are:

$$\phi = (5/5 + \sqrt{5}) \times \sqrt{.5 \times .5}$$

from Jaroslaw Tomasz Wroblewski, and

$$\frac{5/5 + \sqrt{5}}{\log_{\sqrt{5}} 5}$$

from Seth Breidbart of Morgan Stanley & Co., New York.

Phil Hanna gave the following exact solutions for ϕ in eight 8's and nine 9's:

$$\phi = \frac{8 + 8 \times \sqrt{\sqrt{8 + 8} + 8/8}}{8 + 8}$$

$$\phi = \frac{9 + 9 \times \sqrt{\sqrt{9} + 9/9} + 9/9}{9 + 9}$$

Peter Ta-chen Chang asked the following: 1) What is the smallest positive integer that can't be expressed using only four 4's? 2) What is the smallest number of 4's (or some other integer) that will generate all positive numbers? 3) What is the smallest collection of operations that will work with this second question?

Dr. Oz's ϕ contest described in this chapter was stimulated by a paper, published in 1962, in which Conway and Guy asked a similar question for constructing π: J. Conway and M. Guy, "Pi in four 4's," *Eureka* 25 (1962): 18–19.

Another kind of four-4's problem was given by the manufacturer of one of the early hand calculators: Texas Instruments. In *The Great International Math on Keys Book* (Texas Instruments, Inc., 1976, ISBN 0-89512-002-X) the chapter titled "For Four 4's" provided this short exposition:

Here's a brain teaser! Can you (with the help of your calculator, as needed) "build" all the whole numbers between 1 and 100 using only four 4's? Use only the $+$, $-$, $\times$, $/$, $(\,)$, $.$, x^2, $=$, and 4 keys on your calculator. $4! = 4 \times 3 \times 2 \times 1$ is allowed, along with repeating decimal 4 ($\underline{4} = .4444\ldots$).

The first 24 solutions are shown below.

$1 = 4 - 4 + (4/4)$	$2 = (4/4) + (4/4)$	$3 = (4 + 4 + 4)/4$
$4 = 4^2/4 + 4 - 4$	$5 = (4 \times 4 + 4)/4$	$6 = 4 + (4 + 4)/4$
$7 = 4 + 4 - (4/4)$	$8 = 4 + 4 + 4 - 4$	$9 = 4 + 4 + (4/4)$
$10 = (4/\underline{4}) + (4/4)$	$11 = 4^2 - 4 - (4/4)$	$12 = 44 - 4^2 - 4^2$
$13 = 4^2 - 4 + (4/4)$	$14 = 4^2 - (4 + 4)/4$	$15 = (44/4) + 4$
$16 = 44 - 4! - 4$	$17 = 4 \times 4 + (4/4)$	$18 = 4^2 + (4 + 4)/4$
$19 = 4^2 + 4 - (4/4)$	$20 = 4^2 + 4 + 4 - 4$	$21 = 4^2 + 4 + (4/4)$
$22 = 4! - (4 + 4)/4$	$23 = 4^2 + (4! + 4)/4$	$24 = 44 - 4^2 - 4$

For four 4's

Dr. Oz believes it is possible to create integers from 1 to 182, at which point a gap occurs at 183 with no known solution. Other gaps appear at 187, 205, 213, 237, 298, 302, 307, 322, 327, 338, and 339. What is the distribution of gaps for the first 100,000 integers? Dr. Oz doesn't think anyone knows. For more information, see Peter Karsanow, "FAQ for the Four Fours mathematical puzzle," http://www.geocities.com/TimesSquare/Arcade/7810/44sfaq.htm.

48. On Planet Zyph

The top branch does not belong (Figure F48.1). If you superimpose each berry, the black quadrants form a perfect black circle. But in the top fruit, the lower left quadrant is left white! (The phrase "bright light shines through the white portions of the berries" was meant as a hint, but, of course, you could use other creative criteria for removing berries.)

F48.1. Zyph-berry tree solution. (Illustration by Brian Mansfield.)

49. The Jellyfish of Europa

Figure F49.1 shows one solution. Are there any others?

50. Archaeological Dissection

This kind of problem has been known for decades. Cut the cross in the manner shown (Figure F50.1), and the four pieces 1, 2, 3, and 4, will fit together to form a perfect square (Figure F50.2). Cross dissections, such as this one, have intrigued puzzle enthusiasts for at least a century. Here's another one for you to ponder. Can you cut the cross into four identical pieces that can be reassembled to form a square?

F49.1. Solution for the jellyfish of Europa. (Illustration by Brian Mansfield.)

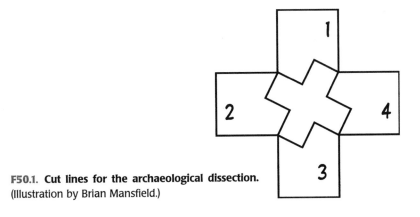

F50.1. Cut lines for the archaeological dissection.
(Illustration by Brian Mansfield.)

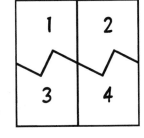

F50.2. Reassembly for the archaeological dissection.
(Illustration by Brian Mansfield.)

51. The Gamma Gambit

Let us suppose that Dr. Oz guesses 1997. This means Dorothy must guess the year before and the year after his guess, namely 1996 and 1998. The odds are significantly in her favor. Try this wager on friends, and let them be amazed, be very amazed. You will almost always win the wager. Try this problem the next time you are practicing transatuaumancy. (If you don't know what this word means, see my book *Dreaming the Future*.)

Can you think of a way to get friends to accept a wager where you get two guesses to their one?

52. Robotic Hand Hive

Here is one solution starting at 1 and ending at 25. Do you think any other solutions exist?

13	12	9	8	7
14	11	10	5	6
15	16	17	4	3
24	23	18	19	2
25	22	21	20	1

53. Ramanujan and the Quattuordecillion

It turns out that it was not too difficult to compute the value for the mysterious nested root. The answer is so simple that it astonishes most colleagues, who expect a noninteger solution. The value of the nested $\heartsuit$ system in Eve's box is simply

$$\heartsuit + 1$$

and this could have been solved decades ago without computers.

In 1911, the 23-year-old Indian clerk named Srinivasa Ramanujan, whom we discussed previously (see Further Exploring section 43), posed the following question (#298) in a new mathematical journal called the *Journal of the Indian Mathematical Society:*

$$? = \sqrt{1 + 2\sqrt{1 + 3\sqrt{1 + \ldots}}}$$

Many months went by, and not a single reader could determine a solution. The difficulty was the infinite nesting of roots. Ramanujan finally threw his hands up and gave the answer, 3. It turns out that he had generated the problem years before in the form of a more general theorem:

$$x + 1 = \sqrt{1 + x\sqrt{1 + (x+1)\sqrt{1 + \cdots}}}$$

Plug in any value you like for x, and the solution is $x + 1$. Plug in $\heartsuit$, and the solution is $\heartsuit + 1$. Try other values and impress your friends.

54. The Lunatic Ferris Wheel

The Ferris wheel, an integral part of large carnivals, was first introduced at the World's Colombian Exposition of 1893 in Chicago. The first Ferris wheel was built by bridgemaker George Ferris, who had an expert knowledge of struts, beams, and supports. This Ferris wheel cost $380,000 to make and stood 264 feet tall.

The mathematics of a Lunatic Ferris Wheel discussed in this chapter comes from Frank A. Farris, "Wheels on Wheels on Wheels – Surprising Symmetry," *Mathematics Magazine* 69(3) (June 1996): 185–9. See also Harold Boas, "Including Maple graphics in LaTeX documents," at http://calclab.math.tamu.edu/~boas/courses/math696/including-Maple-graphics-in-LaTeX.html, and Frank Farris, http://math.scu.edu/~ffarris/homepage.html.

To generate the equations of the seat's motion, recall that a circular orbit can be generated by $x = \cos(t)$, $y = \sin(t)$ as t goes from 0 to 360 degrees or 0 to 2π radians. In fact, this would describe the motion of a chair mounted to a single circle in the Lunatic Ferris Wheel. However, we must add extra terms to account for the combined motion of three wheels. The small wheel is 180 degrees out of phase with the others. This means the path of the chair in Figure 54.1 can be described by

$$x = \cos(t) + \cos(7t)/2 + \sin(17t)/3$$

$$y = \sin(t) + \sin(7t)/2 + \cos(17t)/3$$

That surely would make Dorothy dizzy. (I wonder how many g's of force this would place on her body if Dorothy's seat were traveling at normal Ferris-wheel speeds.) The computational recipe for tracing out the path can be implemented in many software packages. For example, researcher Harold Boas has used Maple software (http://www.maplesoft.com) to plot these curves, using statements resembling these:

```
x := t -> cos(t) + cos(7*t)/2 + sin(17*t)/3;
y := t -> sin(t) + sin(7*t)/2 + cos(17*t)/3;
plot([x(t), y(t), t=0..2*Pi], thickness=2, color=black);
```

The physical interpretation of these formulas involves the position of a particle on a wheel whose center is mounted on the rim of a second wheel whose center is mounted on the rim of a third wheel, each wheel turning at a different rate. Notice that the Ferris wheels can generate trajectories with a variety of other symmetries. For example, if we use −2, 5, and 19 for the various wheel speeds instead of the original 1, 7, and −17, we can generate a figure with sev-

enfold symmetry (Figure F54.1). Which ride would you consider more exciting, the one in Chapter 54's Figure 54.2 or the one shown here? What happens if you add another smaller wheel? Dr. Oz looks forward to seeing your
own Ferris wheel plots for different speeds.

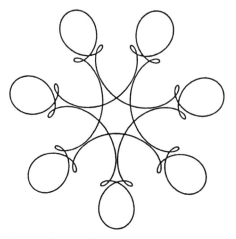

F54.1. The Lunatic Ferris Wheel with wheel speed varied for sevenfold symmetry.
(After Frank Farris, "Wheels on Wheels on Wheels.")

Those of you interested in the deeper mathematics behind these figures
should consult Frank Farris's article. He gives methods for determining the
symmetries of the seat's trajectory once the speeds of each wheel are known.
In this way you can predict the symmetry of Dorothy's path before you even
plot the figure. For those readers knowledgeable with complex mathematical
notation, the curve in Figure 54.2 can be described by

$$F(t) = x(t) + iy(t) = e^{it} + (1/2)e^{7it} + (i/3)e^{-17it}$$

Frank Farris notes that the symmetry of Figure 54.2 arises because 1, 7, and
−17 are all congruent to 1 modulo 6. When t is advanced by one-sixth of $2\pi i$,
each wheel has completed some number of turns, plus one-sixth of an additional turn, resulting in interesting symmetries. The sevenfold symmetry of
Figure F54.1 arises because −2, 5, and 19 are all congruent to 5 modulo 7.

55. The Ultimate Spindle

The graph has gaps because we are plotting real numbers. For real x and y,
$y = x^x$ is only defined if y can be written in the form $(-p/q)$, where p and q
are positive integers and q is odd. This leaves a lot of "holes" in the graph
for $x < 0$, and these holes lie along the curves defined by $y = \pm|x|^x$.

Here's an example that yields a real number.

$$(-2/5)^{-2/5} = \sqrt[5]{25/4} \approx 1.4427$$

We can take odd roots of negative numbers but not even roots. For q odd,

$$\sqrt[q]{x} = -\sqrt[q]{-x}$$

We can take the cube root of -1 to get 1, but we can't take the square root of -1 and get a real number.

We can better understand x^x by examining complex values. The graph of $z = x^x$, where x is still real but z is allowed to be complex, resembles a spindle shape (Figure F55.1). Mark Meyerson of the U.S. Naval Academy in Annapolis, Maryland, one of whose papers describes such spindles, writes: "The word *spindle* is doubly appropriate; not only is the general shape spindlelike, but the graphic consists of a countable infinity of curves or threads wrapped around the shape." To generate the spindle, note that $x^x = e^{x\log x}$ takes the values $e^{x\log|x|+i\pi nx}$. Different threads correspond to different values of n. For a more detailed explanation of the threads and a study of mysterious gaps in the spindle, see Mark Meyerson, "The x^x spindle," *Mathematics Magazine* 69(3) (June 1996): 198–9.

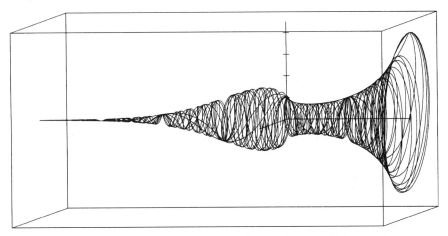

F55.1. The ultimate spindle: the graph of $z = x^x$, when x is real and z is complex. Twenty-one of the threads are plotted for x ranging from −4 to 2. (Figure courtesy of Mark D. Meyerson.)

56. Prairie Artifact

The second pair completes the set because this pair completes every possible pair of the four symbols. Can you justify another answer using other criteria?

57. Alien Pellets

The alien drops eight pellets. He is just multiplying the digits of each number to get the next.

58. The Beauty of Polygon Slicing

In the late 1990s, mathematicians Bjorn Poonen and Michael Rubinstein considered the problem of counting slices in a polygon. The problem is actually quite difficult, far beyond Dorothy's limited capabilities. In a regular n-gon, different numbers of diagonals can meet at an interior point. (Poonen and Rubinstein show that it is impossible to have eight or more diagonals of a regular n-gon meet at a point other than the center.) An explicit description of these diagonal-meeting configurations is very complex. Nonetheless, Poonen and Rubinstein deduced a complicated-looking form of the answer so that we can compute the solution for any n. The number of regions into which the diagonals cut the n-gon is:

$$R(n) = (n^4 - 6n^3 + 23n^2 - 42n + 24)/24$$
$$+ (-5n^3 + 42n^2 - 40n - 48)/48 \cdot \delta_2(n) - (3n/4) \cdot \delta_4(n)$$
$$+ (-53n^2 + 310n)/12 \cdot \delta_6(n) + (49n/2) \cdot \delta_{12}(n) + 32n \cdot \delta_{18}(n)$$
$$+ 19n \cdot \delta_{24}(n) - 36n \cdot \delta_{30}(n) - 50n \cdot \delta_{42}(n) - 190n \cdot \delta_{60}(n)$$
$$- 78n \cdot \delta_{84}(n) - 48n \cdot \delta_{90}(n) - 78n \cdot \delta_{120}(n) - 48n \cdot \delta_{210}(n), \quad \text{for } n > 2$$

$\delta_m(n) = 1$ if $n \equiv 0 \pmod{m}$

        0 otherwise

Isn't that a beauty? This sort of problem has been studied by numerous authors in the 1900s, but it was not until the late 1990s that anyone had a correct formula! Many tried, but all had failed.

For example, the Dutch mathematician Gerrit Bol attempted to give a complete solution in 1938, but several of the coefficients in his formulas were incorrect. Poonen and Rubinstein write, "We relegated much of the work to the computer, whereas Bol had to enumerate the many cases by hand. The task is so formidable that it is amazing to us that Bol was able to complete it, and at the same time not so surprising that it would contain a few errors!" In the late 1950s–early 1960s, mathematicians discovered that no three diagonals meet internally when n is prime.

To answer Dr. Oz's original question, the diagonals of the 30-sided polygon produce 21,480 internal cutouts. For more information, see Bjorn Poonen and Michael Rubinstein, "The number of intersection points made by the diagonals of a regular polygon," *SIAM Journal on Discrete Mathematics* 11(1) (1998): 135–56.

59. Cosmic Call

As Dr. Oz said, at the top of the page are various symbols to help orient any creature receiving the message. At the upper left is the page number in binary. For Figure 59.1, the number is 00001, which means page 1. At the right is a section number, also in binary.

The first page Dr. Oz gave Dorothy is the first page transmitted by Dutil and Dumas. It defines the numbers used in the rest of the transmission. The upper portion of the page defines the numbers 0 through 15 by:

1) a group of dots followed by
2) a binary representation followed by
3) a special symbolic representation.

For example, the upper-right set defines one square dot (■) as a 0001 (in binary) and a symbol resembling

Zeros in the binary numbers are represented by

Zero

A "1" is represented by

One

The choice of these patterns gives the pattern increased resistance against noise in a limited array of dots. At the bottom of the page, we find the primes, starting with a "2" symbol at left:

Symbol for "2"

Continuing from left to right, we find the prime numbers: 2, 3, 5, 7, 11, 13, 17, 19, 23, 29, 31, 37, 41, 43, 47, 53, 59, 61, 67, 71, 73, 79, 83, 89. At the very bottom is the largest prime known to humans at the time the message was beamed into outer space – one measure of a society's intellectual progress:

$$2^{3,021,377} - 1$$

Thus, you now know what Figure 59.1 signifies. As this book goes to press, the largest known prime is

$$2^{13,466,917} - 1$$

This number has more than four million digits and required more than two years and tens of thousands of computers to find.

In Figure 59.2, the physicists begin to show definitions of mathematical operations such as +, −, ×, and ÷. The periodic notation (. . .) is also introduced using fractions.

$1 + 1 = 2$	$1 - 1 = 0$	$1 \times 1 = 1$
$1 + 2 = 3$	$1 - 2 = -1$	$1 \times 2 = 2$
$3 + 2 = 5$	$3 - 2 = 1$	$3 \times 2 = 6$
$4 + 3 = 7$	$4 - 3 = 1$	$4 \times 3 = 12$
$1 + 0 = 1$	$1 - 0 = 1$	$1 \times 0 = 0$

$1/1 = 1$	$1/3 = 0.3333 . . .$
$1/2 = 0.5$	$4/3 = 1.3333 . . .$
$3/2 = 1.5$	$1/9 = 0.1111 . . .$
$1/0 =$ **undetermined**	$2/3 = 0.6666 . . .$
$0/1 = 1$	$1/11 = 0.0909 . . .$
$0 - 1 = -1$	

Figure F59.1 conveys certain chemistry and physics information. The hydrogen atom is displayed with the representation of the proton and the electron. The respective mass and charge of both are also written. The proton's mass is given in relation to that of the electron (proton mass $= 1836 \times$ electron mass). Below this, helium is used to demonstrate the neutron. Ten elements are represented using the protons and neutrons contained in the nucleus of each. For more information and messages, see Stephane Dumas and Yvan Dutil, "SETI," http://www3.sympatico.ca/stephane_dumas/CETI/.

Can you determine what Figures F59.2 and F59.3 signify?

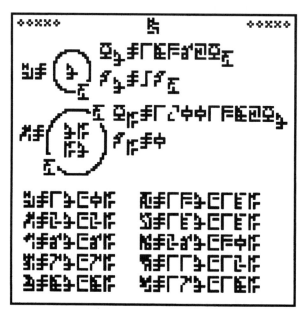

F59.1. Message to the stars. (Courtesy of Yvan Dutil and Stephane Dumas.)

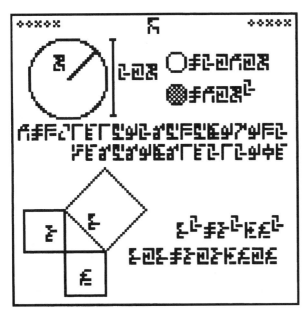

F59.2. Message to the stars. (Courtesy of Yvan Dutil and Stephane Dumas.)

F59.3. Message to the stars. (Courtesy of Yvan Dutil and Stephane Dumas.)

60. Knight Moves

Here is one solution showing a path to the alien. How many other solutions are there? What is the largest sum you can obtain without revisiting squares?

2	3	3	2	1	4
1	2	3	7	2	3
3	2	1	1	3	7
1	1	3	2	3	4
2	2	4	3	4	2
7	4	🕷	3	2	3

61. Sphere

The radius is 18 feet. The area of a sphere is $4\pi r^2$ and the volume is $(4/3)\pi r^3$. According to our problem, this means that both $4r^2$ and $(4/3)r^3$ must be between 1000 and 9999. From the area condition, we get $50 > r > 15$, and from the volume condition we get $20 > r > 9$. Combined, these conditions imply $20 > r > 15$, that is, $r = 16, 17, 18,$ or 19. Next, note that $(4/3)r^3$ can be an integer only if r is divisible by 3. This means $r = 18$ feet. This kind of problem is discussed by Angela Dunn in *Mathematical Bafflers,* but I wonder if it is possible to determine the sphere's radius if the surface area and volume are both five-digit integers times π. How about if the area and volume are both six-digit integers? How about n-digit integers? Can we extend this problem to higher-dimensional spheres?

62. Potawatomi Target

Here is one solution: $12 + 43 + 79 + 16$. Are there others? What is the best strategy you can use to solve this kind of puzzle? One obvious strategy is simply to try all combinations of four numbers; but this is a great deal of work to do by hand because there are $(16, 4) = (16 \times 15 \times 14 \times 13)/(4 \times 3 \times 2 \times 1) = 1820$ possibilities. We could limit the possibilities by separating the numbers into odds and evens: $10, 12, 16, 54, 108$ and $13, 27, 33, \ldots$, and then use the fact that for four numbers to add up to 150, either all four are even, two are even and two are odd, or all four are odd.

63. Sliders

The gray Plex clones are the ones that move to an adjacent square, in the direction shown by the arrows, for one solution to this problem.

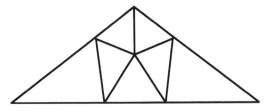

What is the best strategy you can use to solve this kind of puzzle? How much more difficult would this kind of puzzle be to solve if Mr. Plex clones were scattered through a cube rather than square array?

How many other solutions exist?

64. Swapping

Swap the 14 and 15, and the 9 and 12.

65. Triangle Dissection

I learned of this problem from *Martin Gardner's New Mathematical Diversions* (Washington, D.C.: Mathematical Association of America, 1995), 34.

Gardner notes that the problem is exciting because even the best mathematicians are often led astray and come to a wrong conclusion. Most readers sent "proofs" that an obtuse triangle *cannot* be dissected into acute triangles; however, the triangle can in fact be so divided. Figure F65.1 shows a seven-piece pattern that can be adapted to any obtuse triangle. Dr. Oz believes that seven is minimal.

Can you cut an equilateral triangle, square, or regular pentagon into smaller triangles all of which are obtuse or all of which are acute? What is the minimal number of pieces required?

66. A Simple Code

One answer is 0. If the first four cells in a column sum to an even number, the fifth cell in the column is 0, otherwise it is 1. Can you justify other solutions? For example, one colleague gave an answer of 13. Summing the numbers horizontally gives, respectively, 3, 6, 9, and 12. Continuing this trend suggests that the missing number could be 13 (for a sum of 15).

Another colleague got the answer 4, based on the observation that the columns sum to even numbers while the rows sum to multiples of 3 (and all the entries are single digits). What does a person's answer to such an open-ended problem tell you about his or her mathematical or demographic background?

67. Heterosquares

Here is one solution:

9	8	7
2	1	6
3	4	5

Heterosquare

These kinds of squares, called *heterosquares,* caused a flurry of interest in the 1950s. Royal V. Heath, an American magician and puzzle enthusiast, showed that a 2×2 heterosquare formed with the numbers 1, 2, 3, and 4 is impossible. Can you show that other sizes of heterosquares either exist or do not exist?

How many heterosquares are there for a given array size? If you place the numbers at random, what are the odds of producing a heterosquare?

68. Insertion

Dorothy smugly looks at Dr. Oz and says, "$8 + 7 + 6 + 5 + 4 + 3 + 2 + 1 = 36$."

"Pretty good," Dr. Oz says. "But more solutions exist. For example, $87 - 65 - 4 - 3 + 21 = 36$. How many other solutions exist?"

69. Missing Landscape

The missing symbol is . The reason is that, starting from the upper right, and going in a counterclockwise spiral, the following sequence is repeated again and again to form a repetitive skyline:

Isn't that a killer? This is a problem so simple to state, yet so difficult to solve.

Try this one on friends and be assured that none will solve it, even if you offer a $10 reward. One way to get a hint at the answer is to count the number of occurrences of each landscape and observe that one them appears only twice, while the others appear either three or four times.

70. The Choice

There is a *best* strategy, and if Dorothy chooses it, she has at least a $1/e = 0.3678$ chance of picking the Web page with the largest number of words. Here, e is Euler's number, expressed to high precision as:

2.71828182845904523536028747135266249775724709369995957496

(Note that $1/e$ is the limit of the probability of correctly guessing the largest number of words as the number of pages approaches infinity.)

A $1/e$ chance of picking the Web page with the largest number of words is a lot better than most people anticipate. After all, Dorothy could randomly browse to a Web page with the text of a book containing over 60,000 words! Who could guess that? Or she might browse to a Web page with just one or two words if the page contains mostly images. A detailed analysis of the problem is provided in *Martin Gardner's New Mathematical Diversions* (Washington, D.C.: Mathematical Association of America, 1995), 41.

The gist of Gardner's argument, which he obtained from Leo Moser and J. R. Pounder, can be glimpsed in the following analysis. Let n be the number of Web pages and p the number of pages rejected before picking a page with a number of words larger than any on the p pages. Number the Web pages serially from 1 to n. Let $k + 1$ be the number of the Web page bearing the largest number of words. The top number will not be chosen unless $k \geq p$ (otherwise it will be rejected among the first p pages), and then only if the highest number from 1 to k is also the highest number from 1 to p (otherwise this number will be chosen before the top number is reached). The probability of finding the top number in the case in which it is on the $k + 1$ selected Web page is p/k, and the probability that the top number actually is on the $k + 1$ page is $1/n$. Because the largest number can be on only one page, we can write the following formula for the probability of finding it:

$$\frac{p}{n}\left(\frac{1}{p} + \frac{1}{p+1} + \frac{1}{p+2} + \ldots + \frac{1}{n-1}\right)$$

Given a value for n (the number of Web pages), we can determine p (the number to reject) by picking a value for p that gives the greatest value to the above expression. As n approaches infinity, p/n approaches $1/e$. So, a good estimate of p is the nearest integer to n/e. Gardner says that the general strategy, therefore, when the game is played with n Web pages, is to let n/e pages go by, then pick the next number larger than the largest number on the n/e Web pages passed up.

Dorothy's problem assumes that she has no knowledge of the range of numbers on the Web pages and therefore has no basis for knowing whether a single page is high or low with respect to the range of pages and their words.

71. Animal Selection

The set in the upper left

is the answer because it has three different sizes of animal (large, medium, and small). All of the initial examples in the problem have three different sizes.

72. The Skeletal Men of Uranus

The first set, 🕵️🕵️🕵️, is the answer. If we assign 🕵️ = 1, 🕵️ = 2, then each row and column sums to 9. Can you justify a solution with other criteria?

73. Hindbrain Stimulation

The missing symbol is 🐚. There are nine occurrences of each symbol in the array.

74. The Arrays of Absolution

The numbers of the first square total 4, the second 5, the next 6, and so on. Therefore, the second array completes the series.

75. Trochophore Abduction

To be certain that he has two animals of the same species, Dr. Oz must drop four animals – one more than the number of different species. To be certain he has a male–female pair of the same species, he must let drop twelve animals – one more than the total number of animal pairs. Didn't get these answers? Try writing each animal's species and gender on separate scraps of paper. Then put all the papers in a box and withdraw them, one at a time, without looking. Now that you see how it's done, can you think of other "animal and alien" puzzles?

76. The Dream Pyramids of Missouri

Here is one solution: 4321/47,531 = 1/11. What is the best strategy for solving this? Are there other solutions?

77. Mathematical Flower Petal

Here is one possible answer. Can you find others?

10	×	2	+	4	=	24
+	■	+	×	■		+
5	■	10	4	■		2
−	■	+	+	■		+
5	+	5	×	3	=	20
=	■	=	■	=	■	=
10	+	17	+	19	=	46

78. Blood and Water

Believe it or not, both goblets are equally contaminated. The blood contains exactly as much water as the water contains blood. Perhaps the best way to visualize this is to put 6 red stones in a cup (to represent blood) and 6 white stones in another cup (to represent water). Let's assume that your teaspoon holds 3 stones. You take 3 white stones and add them to the cup containing 6 red stones. The blood cup now contains 6 red stones and 3 white stones. Next "stir" the stones in the blood cup. If you dip your teaspoon into the contaminated cup, on average your teaspoon will contain 2 red stones and 1 white stone. Add these to the water cup. Each cup will now have 4 of one stone and 2 of the other. Another explanation: If you moved one teaspoon of blood to the water goblet, you must have moved one teaspoon of water to the blood goblet, because the total amount of liquid in each has not changed.

79. Cavern Problem

Switch the 48 and the 28 cell colors. White cells should be divisible by 3 and gray cells should not.

80. Three Triplets

The answer is "A," because it completes every possible grouping in threes of the four different symbols.

81. Oos and Oob Gambit

Dorothy should take a creature from the flask, pretend it bites her, and immediately drop it to the floor, at which point the creature rapidly scurries away. She then says, "Sorry. Nothing to worry about. We can determine the creature I selected by looking at the one remaining in the flask. If it is an Oos, I must have selected an Oob."

82. Napiform Mathematics

Here is one answer:

6	8	9	7
3	12	5	11
10	1	14	13
16	15	4	2

Antimagic square

J. A. Lindon, an English puzzle aficionado, did much of the pioneering work on "antimagic square" construction. Although there are many ways to create magic squares, there seem to be almost no simple systematic ways for creating antimagic squares. I would be interested in receiving any recipes from readers. It seems that antimagic squares of orders 1, 2, and 3 (i.e., 1×1, 2×2, and 3×3) are impossible, but higher orders occur.

An antimagic square is a special case of a heterosquare (see Chapter 67) for which the sums of rows, columns, and main diagonals form a sequence of consecutive integers. For more information on magic and antimagic squares of all varieties, see my book *The Zen of Magic Squares, Circles, and Stars* (Princeton, N.J.: Princeton University Press, 2002).

83. Toto, Mr. Plex, Elephant

Here is one possible solution. Are there others?

5	6	7	8
4	3	12	9
1	2	11	10

84. Witch Overdrive

Technically speaking, Lissajous curves are a family of curves described by the parametric equations

$$x(t) = A\cos(\omega_x t - \delta_x)$$

$$y(t) = B\cos(\omega_y t - \delta_y)$$

Sometimes these curves are written more simply as

$$x(t) = a\sin(nt + c)$$

$$y(t) = b\sin t$$

Lissajous pronounced (LEE-suh-zhoo) curves are sometimes called Bow-ditch curves, after Nathaniel Bowditch who studied them in 1815. French physicist Jules-Antoine Lissajous (1822–80) researched them in more detail, and the curves have since been applied in physics, astronomy, mathematics, and computer art, and even featured in science-fiction movies. In particular, Lissajous use sounds of different frequencies to vibrate a mirror. A beam of light is reflected from the mirror and traces attractive patterns that depend on the frequencies of the sounds. You can learn more about Lissajous curves at the "MacTutor history of mathematics archive," http://www-groups.dcs.st-andrews.ac.uk/~history/Curves/Lissajous.html, and in Eric Weisstein's *CRC Concise Encyclopedia of Mathematics* (New York: CRC Press, 1999).

Lissajous figures were often used to determine the frequencies of acousti-cal signals. A signal of known frequency was applied to the horizontal axis of an oscilloscope, and the signal to be measured was applied to the vertical axis. The resulting pattern depended on the ratio of the two frequencies.

Back in 1965, ABC-TV used a Lissajous figure for its logo. One of the best-known examples of Lissajous figures was featured in the opening sequence of original *The Outer Limits* TV series, in which the viewer sees a pattern of crisscrossing oscilloscope lines while hearing the dramatic words, "Do not attempt to adjust your picture – *we* are controlling the transmission." You can experiment further with Lissajous figures at Ed Hobbs's "Lissajous lab," http://www.math.com/students/wonders/lissajous/lissajous.html.

The "embellished" Lissajous figures in this book were created by Bob Brill. Figures F84.1 and F84.2 show additional examples. All plots fit neatly into a rectangle whose sides are $2 \times X$ amplitude by $2 \times Y$ amplitude. Although the Lissajous figures described by the computer program are attractive, they are much more so when embellished in various ways. For example, instead of plotting a line segment connecting each point, as outlined in the computer code, you can instead draw a pair of arcs, each of which begins at the cal-culated point. Each arc may be determined by three user-supplied param-eters that specify the angle offset from the current heading, the degree of curvature (e.g., 360 is a circle, 90 a quarter circle, and 1 a straight line), and the length of the straight-line segments composing the arc. One arc may be drawn counterclockwise and offset to the left of the current heading, while the other may be drawn clockwise and offset to the right. Additional exper-iments can be made. For example, you can change the heading before draw-ing the arcs. As each x, y point is calculated, the heading is determined to lie in the direction of the undrawn line connecting the previously calculated point to the newly calculated point. Thus the embellishment follows the path of the curve. We can make the embellishment rotate in various ways by changing the heading. For a more thorough explanation and additional ex-amples, see Bob Brill, "Embellished Lissajous figures," in *The Pattern Book: Fractals, Art, and Nature,* edited by Clifford Pickover (River Edge, N.J.: World Scientific, 1995), 183.

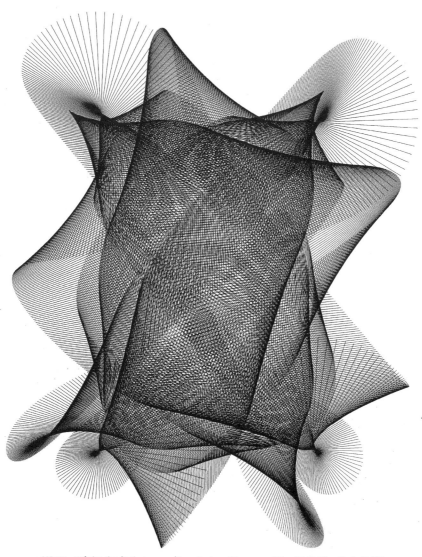

F84.1 **Wicked witch curve.** (Rendering "Dance of the Veils" by Bob Brill.)

85. What Is Art?

The 327849 does not belong. For all others, the sum of all the digits is 30.

86. Wendy Magic Square

All numbers in this magic square are prime. In addition, the primes are consecutive terms in an *arithmetic progression.* In an arithmetic progression each term is equal to the sum of the preceding term and a constant.

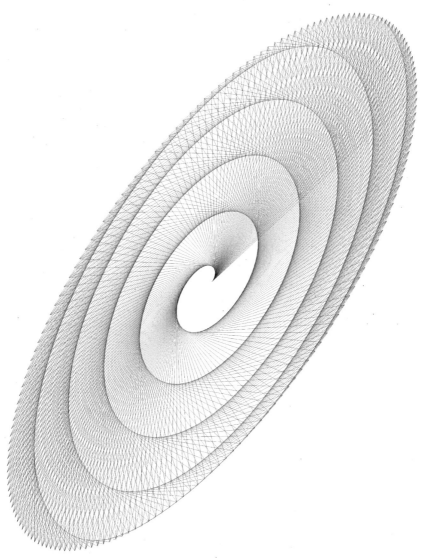

F84.2 Wicked witch curve. (Rendering "Spiralipse" by Bob Brill.)

The magic square has the smallest possible magic constant sum, 3117, for an order-3 square filled with primes in an arithmetic progression. The constant difference is 210.

1669	199	1249
619	1039	1459
829	1879	409

87. Heaven and Hell

This is the longest path of which Dr. Oz is currently aware. It misses three cells. Can you do better? If there is no solution that misses fewer than three cells, how would you prove this? If you were to redesign this maze to make it more difficult, how would you change it?

		20	19	
25	24	21	18	
26	23	22	17	16
27				15
28				14
29				13
2	1	8	9	12
3	4	7	10	11
	5	6		

88. The Stars of Heaven

You can make six squares, represented by the vertices highlighted by circles.

	1	2	3	4	5	6
1		*	O	*		
2	O	*	*	*	*	
3		*		O	*	*
4	*	O	*	*	*	*
5		*	*	*		
6				*	*	*

	1	2	3	4	5	6
1		*	*	O		
2	*	O	*	*	*	
3		*		*	O	*
4	*	*	O	*	*	*
5		*	*	*		
6				*	*	*

	1	2	3	4	5	6
1		*	*	*		
2	*	*	O	*	*	
3		*		*	*	O
4	*	*	*	*	*	*
5		O	*	*		
6				*	O	*

	1	2	3	4	5	6
1		O	*	*		
2	*	*	*	*	O	
3		*		*	*	*
4	O	*	*	*	*	*
5		*	*	O		
6				*	*	*

	1	2	3	4	5	6
1		*	*	*		
2	*	*	*	O	*	
3		O		*	*	*
4	*	*	*	*	O	*
5		*	O	*		
6				*	*	*

	1	2	3	4	5	6
1		*	*	*		
2	*	*	*	*	*	
3		*		*	*	*
4	*	*	*	O	*	O
5		*	*	*		
6				O	*	O

89. Vacation in the Tarantula Nebula

The missing number is 221. Rearrange the digits 123 in every possible way

123 132 213 231 312 321

and then subtract 100 from each result, yielding 023 032 113 131 212 221. Another way to solve the problem is to exchange the second and third digits of every other number to get the next number. Thus, 023 becomes 032; 113 becomes 131; and 212 becomes 221. Of course, problems like this may yield other equally valid solutions, and I would be interested in hearing how readers justify other solutions. One interesting avenue of future research involves the cataloging of all the solutions that people can justify. Is my solution the most commonly given solution? What might we learn about how people think from such studies? Would we find that peoples' solutions depend on age, gender, occupation, or culture?

90. Hot Lava

Please pardon Dr. Oz for this gruesome puzzle. Uncle Henry's, Aunt Em's, and Dorothy's odds of "winning" are 1/2, 1/4, and 1/8. Bet on Uncle Henry!

Because Uncle Henry goes first, he actually has twice the chance that Aunt Em does. Because Aunt Em goes second, she has twice the chance that Dorothy has. Therefore their chances are in a ratio of 4 to 2 to 1. For example, in order for Aunt Em to win, we must have Uncle Henry die (1/2) and Aunt Em survive (1/2) giving a combined probability of 1/4. For Dorothy to win, we must have Uncle Henry die (1/2), Aunt Em die (1/2), and Dorothy survive (1/2) giving a combined probability of 1/8. The chance that no one will win is the same as the chance Dorothy will win. Do you see why? Chapter 19, "The Mystery of Phasers," gave a more complex problem involving similar kinds of probabilities.

If you want a more challenging problem, how would your answer change if Uncle Henry, Aunt Em, and Dorothy continue playing until someone dies? (In this morbid variation, the winner is the first person who dies.) After Dorothy tries, Uncle Henry tries again, and they continue, in the same order, until someone "wins." What are the odds for each person winning? How can you justify your solution?

The solution for this devilish new problem is as follows:

UNCLE HENRY: $57.14\% = 50 + 50/8 + 50/8^2 + 50/8^3 + 50/8^4 + 50/8^5 \ldots$

AUNT EM: $28.57\% = 25 + 25/8 + 25/8^2 + 25/8^3 + 25/8^4 + 25/8^5 \ldots$

DOROTHY: $14.28\% = 12.5 + 12.5/8 + 12.5/8^2 + 12.5/8^3 + 12.5/8^4 + 12.5/8^5 \ldots$

For example, Uncle Henry's chances of winning on his first turn are 50%, and his cumulative chances diminish on each subsequent round across the three players by 2^3 or 8. Notice that because Uncle Henry goes first and then

Aunt Em followed by Dorothy, Aunt Em will have half the cumulative chance of Uncle Henry, and Dorothy will have half the cumulative chance of Aunt Em of dying (i.e., winning). As they keep playing, their cumulative chances added together approach 100% for the three people.

91. Circular Primes

Various mathematicians have made observations regarding circular primes. See, for example, Patrick De Geest's "World of Numbers," http://www.ping.be/~ping6758/menu1.shtml and http://www.worldofnumbers.com/index.html. Also see Keith Devlin's *All the Math That's Fit to Print*.

If we consider multidigit numbers, a circular prime can contain only the digits 1, 3, 7, and 9. Note that a circular prime can never contain an even digit, because once such a digit is positioned at the end of the number, the number is composite (nonprime). If a number ends with 5, the number is divisible by 5. Circular primes are difficult to find as we consider large numbers! Keith Devlin in *All the Math That's Fit to Print* comments on the paucity of a related set of primes called *permutable primes:*

Permutable primes are prime numbers that remain prime when you rearrange their digits in any order you please. For example, 13 is a permutable prime, since both 13 and 31 are prime. Again, 113 is a permutable prime, since it and each of the numbers 131 and 311 is prime. It is known that there are only seven such numbers within reasonable range (less than about 4 followed by 467 zeros, in fact). You now know two of them. Find the other five.

Other authors have called numbers in which every permutation of digits is a prime, *absolute primes*. In the 1970s, the following absolute primes were reported: 2, 3, 5, 7, 11, 13, 17, 31, 37, 71, 73, 79, 97, 113, 131, 199, 311, 337, 373, 733, 919, 991, 11111111111111111111, and 1111111111111111111111. For more information on absolute primes, see T. N. Bhargava and P. H. Doyle, "On the existence of absolute primes," *Mathematics Magazine* 47 (1974), 233.

One of my favorite bizarre questions involving prime numbers relates to the wonderful *special augmented primes*. You can augment a prime simply by placing a 1 before and after the number. The augmented prime is "special" if it yields an integer when divided by the original prime number. For example, 137 is such a number because 137 is prime and because 11,371/137 yields an integer, namely 83. Similarly, 9091, 909,091 and 5,882,353 are also numbers of this kind. Are there other such numbers? How rare are special augmented primes?

Finally, consider *obstinate numbers*. In 1848, Camille Armand Jules Marie, better known as the "Prince de Polignac," conjectured that every odd number is the sum of a power of 2 and a prime. (For example, $13 = 2^3 + 5$.) He claimed to have proved this to be true for all numbers up to three million, but de Polignac probably would have kicked himself if he had known that he missed 127, which leaves residuals of 125, 123, 119, 111, 95, and 63 (all composites) when the possible powers of 2 are subtracted from it. There are another sixteen of these odd numbers – which my colleague Andy Edwards calls "obstinate numbers" – that are less than 1000. There are an infinity of obstinate numbers greater than 1000. Most obstinate numbers we have discovered are prime themselves. The first composite obstinate number is 905. For more information, see David Wells, *The Penguin Dictionary of Curious and Interesting Numbers* (New York: Penguin, 1986), 136–7. Also see Albert Beiler, *Recreations in the Theory of Numbers* (New York: Dover, 1966), 226.

Prince de Polignac (1832–1913) was born in Millemont, Seine-et-Oise, France, and as a young man won first prize in a Europe-wide mathematics contest. In 1953, he served in the Crimean War, after which he traveled to Central America to explore and study plants. In 1861, during America's Civil War, de Polignac offered his services to the American Confederacy, and he was eventually promoted to major general.

92. The Truth about Cats and Dogs

Here is one solution. How many others can you find?

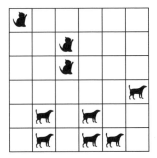

What is the best strategy for solving this? Extend the puzzle to a 3-D array, such as a cubical array of cats and dogs. Better yet, present this kind of problem to friends on a hypercube, a four-dimensional analogue of a cube.

93. Disc Mania

Here is one solution:

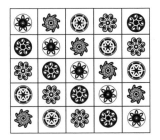

In other words, switch 1 and 2; 3 and 4; and 5 and 6.

1		2		
			3	
	5	6	4	

94. $n^2 + m^2 = s$

On average, the number of ways of expressing a positive integer s as a sum of two integral squares, $n^2 + m^2 = s$, is π. The following code is Harry J. Smith's outline of a computer recipe for empirically investigating the problem.

```
s_Max   : Max s used for current estimate
n       : n used in s = n*n + m*m
m       : m used in s = n*n + m*m
n2      : n squared
m2      : m squared
s       : Sum of two squares = n*n + m*m
t       : t = total number of ways integers <= s_Max can be
            written as the sum of two squares
a       : Average number of ways, t / s_Max;
Error   : Computed error = Pi - a
Done    : Boolean done flag

s_Max ← 1;
repeat
  n ← 0;  t ← 0;
  repeat
    m ← -1;  n ← n + 1;  n2 ← n * n;  Done ← True;
    repeat
      m ← m + 1;  m2 ← m * m;  s ← n2 + m2;
      f (s <= s_Max) {
        Done ← False;
        if (m = 0) or (m = n)
          then  t ← t + 4
          else  t ← t + 8;
      }
      else
          m ← n;
    until (m = n);
  until Done or (interrupted by operator);
  a ← t / s_Max;  Error ← Pi - a;
  output a, Error, s_Max;
  s_Max ← 2 * s_Max;
until (interrupted by operator);
```

Notice that if a solution exists, then when $m = n$ there are 4 ways to form a number: (n, n), $(n, -n)$, $(-n, n)$, $(-n, -n)$. When $m = 0$ there are 4 ways for each entry: $(0, n)$, $(0, -n)$, $(n, 0)$, $(-n, 0)$. When $m > 0$ and $m \neq n$ there are 8 ways for each entry: (m, n), $(m, -n)$, $(-m, n)$, $(-m, -n)$, (n, m), $(n, -m)$, $(-n, m)$, $(-n, -m)$. If we scan the results up to 128 (i.e., s_Max = 128 in the code), we find 8 entries with $m = n$, 11 entries with $m = 0$, and 41 entries with $m \neq n$. Harry Smith notes that the estimate for π then is $[4 \times (8 + 11) + 8 \times 41]/128 = 3.15625$. Try calculating this for larger values of s_Max.

Here's another fascinating piece of π trivia. The probability that any two randomly chosen integers don't have a common divisor – as, for example, 4 and 9 do not, but 6 and 9 do since they are both divisible by 3 – is $6/\pi^2$. Why in the world does π unexpectedly crop up here?

95. 2, 271, 2718281

The following numbers, sometimes known as e-primes, are the primes so far discovered in the decimal expansion of the transcendental number e:

2

271

2718281

271828182845904523536028747135266249775724709369999595749669676277240766303535475945 71

In other words, the first digit of e is examined, then the first two digits, then the first three digits, and so on, until a prime number is discovered. You can search for these kinds of numbers using the software package Maple and implementing the following code:

```
Digits:=110; n0:=evalf(E); for i from 1 to 100 do t1:=trunc(10^i*n0); if
isprime(t1) then print(t1); fi; od:Keywords: base,nonn,huge.
```

Note that the transcendental number e can be defined as the sum of a series in which the series terms are the reciprocals of the factorial numbers: $e = 1/0! + 1/1! + 1/2! + \ldots = 2.7182818284590\ldots$. Similar sequences can be found at the Web site "Sloane's on-line encyclopedia of integer sequences," http://www.research.att.com/~njas/sequences/.

The number e, like π, goes on and on with no obvious patterns to the digits visible through casual inspection. For e aficionados, here are the first few hundred digits:

2.718281828459045235360287471352662497757247093699959574
9669676277240766303535475945713821785251664274274663919 3
2003059921817413596629043572900334295260595630738132328
6279434907632338298807531952510190115738341879307021540 89
1499348841675092447614606680822648001684774118537423454 4
2437107539077744992069551702761838606261331384583000752 0
4493382656029760673711320070932870912744374704723069697

720931014169283681902551510865746377211125238978442505695
369677078544996996794686445490598793163688923009879312 7
736178215424999229576351482208269895193668033182528869 39
849646510582093923982948879332036250944311730123819706 8
416140397019837679320683282376464804295311802328782509 81
945581530175671736133206981125099618188159304169035159888
851934580727386673858942287922849989208680582574927961 0
484198444363463244968487560233624827041978623209002160 9
902353043699418491463140934317381436405462531520961836 90
888707016768396424378140592714563549061303107208510383 75
051011574770417189861068739696552126715468895703503540212
340784981933432106817012100562788023519303322474501585390
473041995777709350366041699732972508868769664035557071 6
226844716256079882651787134195124665201030592123667719432
527867539855894489697096409754591856956380236370162112 0
477427228364896134225164450781824423529486363721417402 38
893441247963574370263755294448337998016125492278509257 7
825620926226483262779333865664816277251640191059004916 44
998289315056604725802 77

(Please forgive Dr. Oz for listing so many digits. He just loves staring at them.)

Kevin S. Brown has conducted fascinating research looking for possible patterns in these digits strings (for example, Brown checks for "normality" of the digits). These experiments are reported at: Kevin S. Brown, "Is *e* normal?" http://www.mathpages.com/home/inumber.htm.

Jason Earls has discovered the largest known $\sqrt{2}$ prime in the decimal expansion of the square root of 2 (1.4142 . . .):

14142135623730950488016887242096980785696718 75376948073

96. Android Watch

Figure F96.1 shows one solution. Are there others? Notice that in this solution no robot obscures the vision of another. They can't even see one another in this solution. What is the maximum number of androids you can place at intersections so that the androids cannot see one another? Given four androids stationed at intersections, what arrangement can you create such that the *least* number of runways are observed by the androids?

Your next task is to design a surveillance network covering all the runways with the fewest number of androids so that each android is in view of at least one other android. After you have achieved this goal, constrain the number of visible robots to two. This would form a surveillance "ring," with a fault-detection mechanism similar to those used in token-passing rings in computer networks.

F96.1. Androids placed at intersections. (Illustration by Brian Mansfield.)

97. Knight Moves

Figure F97.1 shows one solution that produces the sum $1 + 3 + 5 + 6 + 8 + 9 + 11 + 12 + 10 + 11 + 13 = 89$. How many others can you find?

Perhaps an even more difficult problem is to find a path from bottom to top that sums to 101. Can you find a solution that passes through the special "zero" circle in the middle to solve the problem?

F97.1. One solution to More Knight Moves. (Illustration by Brian Mansfield.)

(If you pass through this space, nothing is added to your sum when you land on this location.) How many different moves can a knight make on this board without landing on a circle twice?

98. Pool Table Gambit

Figure F98.1 shows one solution. Are there others?

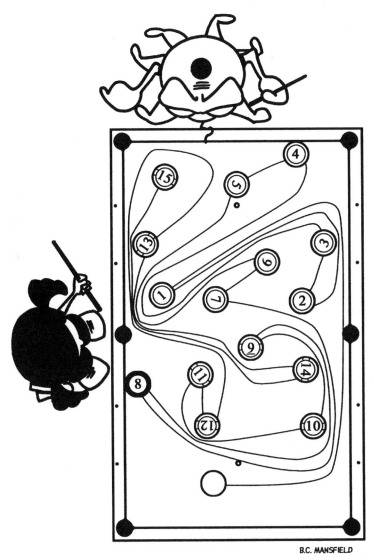

B.C. MANSFIELD

F98.1. Pool Table Gambit solution. (Illustration by Brian Mansfield.)

99. A Connection between π and e

The next number in the bizarre sequence 6, 28, 241, 11,706, 28,024, 33,789, 1,526,800 – sometimes known as Pickover's sequence – is 73,154,827. The numbers count the position of the first occurrence of the starting digits in e within π. For example, 2 occurs in the sixth place of 3.1415926. No one has yet discovered the next number in the sequence.

1st Digits of *e*	Position in π
2	6
27	28
271	241
2,718	11,706
27,182	28,024
271,828	33,789
2,718,281	1,526,800
27,182,818	73,154,827
271,828,182	?

The largest known consecutive digit string in *e* and also in π is 307381323, which is found in *e* here:

2.71828182845904523536028747135266249775724709369995957496696762772407663035354759457138217852516642742746639193200305992181741359662904357290033429526059563073813232862794349076323382988075319525101901157383418793070215408914 99348841

This string is also found at position 29,932,919 in π, counting from the first digit after the decimal point in π. The string and surrounding digits in π are

72064906369568307473307381323845892960611161408236

For those of you who wish to do further searches in π, see: Dave Andersen, "The Pi Search Page," http://www.angio.net/pi/piquery.

Assuming that π and *e* are essentially infinite, "patternless" sequences of digits, then both numbers should have arbitrarily large chunks of digits in common. In fact, given any particular sequence of 1000 digits (for example, the first 1000 digits of π), this sequence is guaranteed to exist somewhere in *e*. However, the process of locating such digit commonalties may prove difficult in practice.

Here's another odd sequence: the Earls sequence, named after its researcher, Jason Earls. The sequence catalogs the first occurrence of *n n*'s in the decimal expansion of π. For example, 1 occurs in position 1 after the 3 and the decimal point. (We don't count the initial 3 when discussing positions in π.) Two 2's or 22 occurs at position 135. Three 3's or 333 occurs at position 1698. The sequence grows quickly: 1, 135, 1698, 54,525, 24,466, 252,499, 3,346,228, 46,663,520 . . . . Note that *999999999*, or nine 9's, does not occur anywhere in π's first 100,000,000 digits. Earls also found the longest known smoothly undulating numbers in π: 242424242 at position 242,421 and 292929292 at position 69,597,703. (Numbers are smoothly undulating if two digits alternate.) Is it coincidence that the first undulating number and its position have obvious similarities? The largest known Fibonacci number found in π

is 39088169, located at position 36,978,613. The largest known consecutive string of even numbers starting with 2 – that is, 2, 4, 6, 8, 10, which in π may be represented as 0204060810 – was found at position 78,672,424. The string and surrounding digits are

20596116031919639159020406081075649735479852708849

100. Venusian Number Bush

Figure F100.1 shows one solution. Are there others?

B.C. MANSFIELD

F100.1. Solution to Venusian Number Bush. (Illustration by Brian Mansfield.)

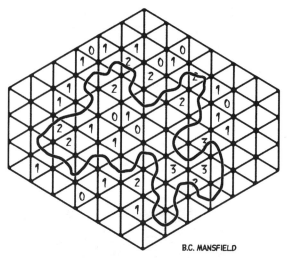

F101.1. Triangle Cave, one solution. (Illustration by Brian Mansfield.)

101. Triangle Cave

Figure F101.1 shows one solution. How many others can you find? Can you find a solution that encloses a bigger area than shown here?

102. Rat Attack

Figure F102.1 shows a good way of cutting through the walls. You can see that only nine holes are required to get the rat from top to bottom. Stay away from the rats! Can you find other solutions that require fewer walls to traverse?

F102.1. Solution to Rat Attack. (Illustration by Brian Mansfield.)

103. The Scarecrow Formula

Because an isosceles triangle has two equal sides, what the Scarecrow said,

"The sum of the square roots of any two sides of an isosceles triangle is equal to the square root of the remaining side."

might be represented mathematically as

$$\sqrt{a} + \sqrt{a} = \sqrt{c}$$

Under what circumstances, if any, could this equation hold true? Obviously, what the scarecrow meant to say was, "The square of the hypotenuse of a *right* triangle is equal to the sum of the squares of the remaining two sides."

With minimal manipulation, we find that $\sqrt{a} + \sqrt{a} = \sqrt{c}$ reduces to $2\sqrt{a} = \sqrt{c}$, or $c = 4a$. However, it is not possible for the unequal side of an isosceles triangle to be four times as long as another side. Test this for yourself by trying to draw such a triangle.

However, perhaps what the Scarecrow also meant was $\sqrt{a} + \sqrt{c} = \sqrt{c}$. However, this formula implies $a = 0$, which is also not possible for a triangle.

My friend Andres Delgado believes that the author of the Oz movie screenplay intentionally had the Scarecrow speak an impressive-sounding formula that was mathematical gibberish. Because Oz gives the Tin Man a "fake" heart (a heart-shaped watch that ticks) and the Cowardly Lion "fake" courage (in the form of an attractive medal), it would make sense that the Scarecrow's impressive-sounding formula is sufficient to give both him and his friends renewed confidence in his ability whether or not the formula is correct. In any case, perhaps the Scarecrow's formula is valid in some kind of warped space in which a line is not the shortest path between two points, and perhaps the land of Oz is based on some strange geometry.

Note that Homer J. Simpson in the fifth season (1993–4) of the TV show *The Simpsons* repeats the Scarecrow's line.

104. Circle Mathematics

We may define the *curvature* of a circle as the reciprocal of the circle's radius. Hence, if a circle is one-half the size of another (here, the outermost) circle, its curvature is twice that of the larger circle. This corresponds to the two circles labeled with a "2" in the figure. The next two smallest circles that fit in the remaining space between the "2" circles each have a radius of $1/3$ (compared to the outermost circle) and a curvature of 3. Once we have the values for the three largest circles, we can compute the radii of the others using a formula by philosopher and mathematician René Descartes (1596–1650). Given four mutually tangent circles with curvatures a, b, c, and d, the Cartesian circle equation specifies that $(a^2 + b^2 + c^2 + d^2) = \frac{1}{2}(a + b + c + d)^2$. (The outside circle in the figure is given a curvature of -1 relative to the inside circles; the negative sign indicates the other circles touch the large from the inside rather than the outside.) If the curvatures of the three initial circles are integers (the largest bounding circle and two inner circles that fit side by side, in this case the two 2 circles) , the curvature of every smaller circle is also an integer. In 2001, mathematician Allan R. Wilks of AT&T Laboratories discovered that the centers (x, y) of the circles are all rational numbers (fractions) – if the first circle is plotted so that its center is at (0, 0). Additionally, Cx and Cy are integers, where C is the curvature of a circle whose origin is located at (x, y). This circle research is an excellent example of how scientists sometimes start with a graph or visualization and then discover interesting results when trying to understand the patterns. For more information, see Ivars Peterson, "Circle game," *Science News* 159(16) (April 21, 2001): 254–5.

Figure F104.1 and F104.2 show how circles can be packed together in a variety of ways. For example, Figure 104.1 depicts a symmetrical set of circles packed between parallel lines. Starting at the right "1" circle and following its tail to the left, we find a sequence of perfect square numbers: 1, 4, 9, 16, 25, 36, ... Can you find any other interesting patterns of numbers?

Figure 104.2 shows a nonsymmetrical packing. Note that these kinds of structures are fractal and exhibit self-similarity. There is an infinite universe to explore as one continues to "magnify" regions of the pattern.

105. *A, AB, ABA*

Just for kicks, let's choose two small numbers for A and B, such as $A = 2$ and $B = 3$. Therefore we want to find the smallest integer n such that, when divided by 2, 232, and 2323, the remainders are 2, 23, 232, respectively. This number n is 197,687 because

$$197{,}687 \div 23 \ \ = 8595 + 2/23$$

$$197{,}687 \div 232 \ = 852 + 23/232$$

$$197{,}687 \div 2323 = 85 + 232/2323$$

Can you prove that $A = 2$ and $B = 3$ produce the smallest integer n?

Tim Petersen explains that Dorothy is attempting to find the smallest integer n such that:

I) $n = 23a + 2$

II) $n = 232b + 23$

III) $n = 2323c + 232$

where $a, b,$ and c are themselves integers. Since III will produce the largest n, it may be intuitive to try to minimize III while satisfying equations I & II. Combining III and I gives: $a = (2323c + 230)/23 = 101c + 10$. This means that any integral choice of c in III will still satisfy I. Also, combining III and II gives $b = (2323c + 209)/232 = 10 + (3c + 209)/232$, and $b - 11 = (3c - 23)/232$. For $b - 11$ to be an integer, $3c - 23$ must be $232m$, where m is any integer.

F104.1. Symmetrical packing of circles between two lines. (Figure courtesy of Allan R. Wilks, AT&T Labs – Research.)

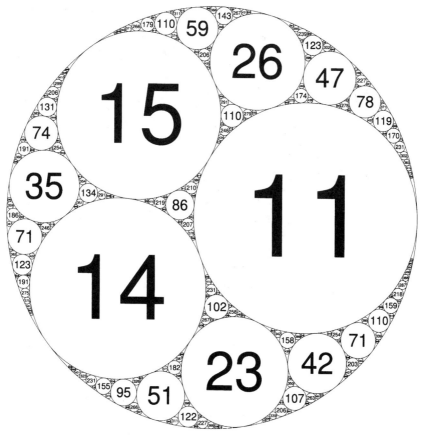

F104.2. Nonsymmetrical packing of circles. (Figure courtesy of Allan R. Wilks, AT&T Labs – Research.)

Thus $c = (232m + 23)/3$. Minimizing both b and c, whilst simultaneously satisfying equation I, it is apparent that m must be 1 and cannot be zero. This gives $c = 85$, which implies by III that $n = 197{,}687$. In 1976 the United States celebrated its Bicentennial year.

Tim Petersen has done significant research in the area of "*ABA*" numbers. He writes:

I spent another 23 minutes today on deriving a general method for producing the undulating remainders. A little algebra shows that for a favorite number *AB* (e.g., for 23, $A = 2$, $B = 3$), n need only be of the form

$$n = [101(10A + B)(111A + 11B) \div B] + 110A + B$$

Thus, if 45 is your favorite number, then $n = 454{,}036$ will produce three of the desired remainder undulations. Note that my equation will not work for certain values of *A* and *B*. Can you tell which ones?

106. Ants and Cheese

Figure F106.1 shows one solution. Are there others?

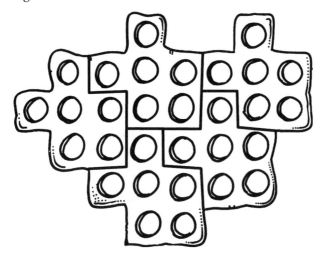

F106.1. The cheese divided into five identical pieces.

107. The Omega Crystal

The volumes of the cubes that form the Omega Crystal are part of this series:

$$1 + \frac{1}{2\sqrt{2}} + \frac{1}{3\sqrt{3}} + \frac{1}{4\sqrt{4}} + \ldots + \frac{1}{n\sqrt{n}} + \ldots$$

For example, if our units of measurements are feet, the first box would have a volume of one cubic foot and the next box would have a volume of about 0.35 cubic feet. This series converges. The total volume of the Omega Crystal is thus finite, but the surface area is infinite! Of course, an infinite object such as this cannot truly be constructed because the boxes would eventually become smaller than an atom, but this is a wonderful example of a wide class of mathematical objects that have finite volumes but infinite surface areas.

By summing the series on a computer, we find the series converges to a value of around 2.61. For those highbrow mathematicians among you, note that the series converges to Riemann's zeta function, $\zeta(3/2)$. For computer programmers, try this C program:

```c
#include <stdio.h>
#include <math.h>
main(int argc, char * arg[ ])
{
    long int size = 5000;
    long int n;
    double d;
    double sum = 0;
    if (argc > 1)
        size = atol(argv[1]);
    for (n = 1; n <= size; n++)
        { d = n;
            sum = sum + 1 / (d * sqrt(d));
        }
    n--;
    printf("Sum with %ld terms is %g.\n",n,sum);
}
```

108. Attack of the Undulating Undecamorphs

I know of no other undulating undecamorphic integers. Perhaps one such beast exists, but no one knows for sure. I discuss all facets of polymorphic numbers and undulating polymorphic numbers in my book *Computers and the Imagination* (New York: St. Martin's Press, 1992).

While at least one multidigit polymorphic integer may exist for all polygonal numbers, this is still an unsolved question, and I invite you to prove or disprove this conjecture. Undulating polymorphic integers are much rarer. In *Computers and the Imagination,* I searched for undulating polymorphic integers $2 < n < 100$, scanning all ranks less than 10,000. The record-holding undulating polymorph (for $2 < n < 100$) is the 52-gon, which has 15 undulating occurrences:

r	p(r)
160	636,**160**
240	1,434,**240**
265	1,749,**265**
281	1,967,**281**
376	3,525,**376**
401	4,010,**401**
480	5,748,**480**
505	6,363,**505**
560	7,826,**560**
616	9,471,**616**
801	16,020,**801**
856	18,297,**856**
1201	36,03**1,201**
3601	324,093,**601**
6576	1,080,93**6,576**

All known undulating 52-morphic integers of rank < 10,000

r	p(r)	r	p(r)
1	**1***	80	79,0**80***
4	15**4***	81	81,0**81**
5	25**5**	88	95,7**88**
8	70**8***	89	97,9**89***
9	90**9***	96	114,0**96**
16	30**16**	97	116,4**97**
17	34**17***	145	261,**145**
24	69**24***	160	318,**160***†
25	75**25**	161	322,**161**†
32	12,4**32**	176	385,**176**
33	13,2**33**	225	630,**225**
40	19,5**40**	240	717,**240***†
41	20,5**41**	241	723,**241***†
48	28,2**48**	256	816,**256**
49	29,4**49**	305	1,159,**305**
56	38,5**56**	320	1,276,**320**†
57	39,9**57**	321	1,284,**321**†
64	50,4**64**	336	1,407,**336**
65	52,0**65**	385	1,848,**385***
72	63,9**72**	400	1,995,**400**†
73	65,7**73**	401	2,005,**401**†
		416	2,158,**416**

* denotes undulation;
† denote three-digit twin 27-morphs.

Searching for undulating 27-morphic integers

For those of you completely enamored with polymorphic integers, please feel free to consider the wonderful *twin 27-morphic integers.* I use the term *twin polymorphic integers* for polymorphic integers with consecutive rank values. For example, the 27-morphic integers follow certain notable patterns (see table, above right). Those numbers with two digits exhibit twinning with a separation distance of 7 between successive twins, for example, (16, 17), (24, 25), (32, 33), ... The three-digit undulating 27-morphic integers also periodically exhibit a twinning, and the last digits repeat the patterns 5 0 1 6, 5 0 1 6, ... I challenge you to find any other examples of twin polymorphic numbers for $n \neq 27$.

Further Reading

Brill, Bob (1995). "Embellished Lissajous figures," in *The Pattern Book: Fractals, Art, and Nature,* edited by Clifford Pickover. River Edge, N.J.: World Scientific, 183.

Caldwell, Chris K. "Primorial and factorial primes."
http://www.utm.edu/research/primes/lists/top20/PrimorialFactorial.html

De Geest, Patrick. "World of numbers."
http://www.ping.be/~ping6758/   and
http://www.worldofnumbers.com/index.html

Dumas, Stephane, and Yvan Dutil, "SETI"
http://www3.sympatico.ca/stephane_dumas/CETI/

Dunn, Angela. (1980). *Mathematical Bafflers.* New York: Dover.

Farris, Frank A. (1996) "Wheels on wheels on wheels – surprising symmetry," *Mathematics Magazine 69*(3) (June): 185–9.

Gardner, Martin (1995). *Martin Gardner's New Mathematical Diversions.* Washington, D.C., Mathematical Association of America.

Honsberger, Ross (1985). *Mathematical Gems III.* New York: Mathematical Association of America.

Jörgenson, Loki. "Visible structures in number theory."
http://www.cecm.sfu.ca/~loki/

Kestenbaum, David (1998). "Gentle force of entropy bridges disciplines," *Science* 279: 1849.

Meyerson, Mark (1996). "The x^x spindle," *Mathematics Magazine* 69(3) (June): 198–9.

Peterson, Ivars (2000). "The power of partitions," *Science News* 157(25) (June 17): 396–7.

Pickover, Clifford (1992). *Computers and the Imagination*. New York: Wiley.

Pickover, Clifford (1992). *Mazes for the Mind*. New York: Wiley.

Pickover, Clifford (1995). *Keys to Infinity*. New York: Wiley.

Pickover, Clifford (2000). *Wonders of Numbers*. New York: Oxford University Press.

Pickover, Clifford (2001). *The Alien IQ Test*. New York: Dover.

Pickover, Clifford (2002). *Mind-Bending Puzzle Calendar*. Rohnert Park, California: Pomegranate.

Poonen, Bjorn, and Michael Rubinstein (1998). "The number of intersection points made by the diagonals of a regular polygon," *SIAM Journal on Discrete Mathematics* 11(1): 135–56.

Sery, Robert S. (1999–2000). "Prime-poor equations of the form $i = x^2 - x + c$, c odd," *Journal of Recreational Mathematics* 30(1): 36–40.

Thinkquest, Inc. "Expected Value of a Random Variable." http://library.thinkquest.org/10030/5rvevoar.htm

Thinkquest, Inc. "Interview: Dr. Yvan Dutil on Astrobiology." http://library.thinkquest.org/C003763/index.php?page=interview01

Trigg, Charles (1985). *Mathematical Quickies*. New York: Dover.

Index